THE MATERNAL GENETIC LINEAGES OF ASHKENAZIC JEWS

....

Kevin Alan Brook

THE MATERNAL GENETIC LINEAGES OF ASHKENAZIC JEWS

Kevin Alan Brook

BOSTON
2022

Library of Congress Cataloging-in-Publication Data

Names: Brook, Kevin Alan, author.
Title: The maternal genetic lineages of Ashkenazic Jews / Kevin Alan Brook.

Description: Boston : Academic Studies Press, 2022. | Includes
  bibliographical references and index.
Identifiers: LCCN 2022021565 (print) | LCCN 2022021566 (ebook) | ISBN
  9781644699836 (hardback) | ISBN 9781644699843 (paperback) | ISBN
  9781644699850 (adobe pdf) | ISBN 9781644699867 (epub)
Subjects: LCSH: Jews–Origin. | Jews--genetics. |
  Jews–Europe–Identity–History. | Europe--Ethnic relations–History.
Classification: LCC GN547 .B86 2022 (print) | LCC GN547 (ebook) | DDC
  940/.04924--dc23/eng/20220610
LC record available at https://lccn.loc.gov/2022021565
LC ebook record available at https://lccn.loc.gov/2022021566

ISBN 9781644699836 (hardback)
ISBN 9781644699843 (paperback)
ISBN 9781644699850 (adobe pdf)
ISBN 9781644699867 (epub)

Book design by Lapiz Digital Services
Cover design by Ivan Grave

Published by Academic Studies Press
1577 Beacon Street
Brookline, MA 02446, USA
press@academicstudiespress.com
www.academicstudiespress.com

This book is dedicated to my nineteenth-century Ashkenazic Jewish ancestors, including Sosie Adler from Ukraine (haplogroup H7j1), Lena Levy from Belarus (haplogroup K1a9), and Chana Maria Kronfeld Dzięciołowska from Poland (haplogroup K1a1b1a).

Contents

Summary

This book presents up-to-date information on the origins of the Ashkenazic Jewish populations of Germany and Poland, and their millions of descendants around the world, based on genetic research on modern and pre-modern populations. It focuses on the 129 maternal haplogroups that the author confirmed that Ashkenazim have acquired from distinct female ancestors who were indigenous to diverse lands that include Israel, Italy, Poland, Germany, North Africa, and China, revealing their Israelite inheritance and the lasting legacy of conversions to Judaism. Only some of these haplogroups were part of the Ashkenazic founding population in early medieval Germany.

Some of the findings appear in print for the first time here, such as those for the author's haplogroup H7j1. The presences of the maternal haplogroups H2a1e1a and U5a1d2b in the Ashkenazic population are new discoveries that the author made while writing this book.

Additionally, major Ashkenazic paternal lineages are summarized. The ancestors of Ashkenazic male lineages were largely of West Asian origin, including many from the Israelites, but some were of European origin and one was from Alania in the Caucasus region.

Genetic connections between Ashkenazic Jews and non-Ashkenazic Jewish populations (Turkish Jews, Moroccan Jews, Tunisian Jews, Iranian Jews, Cochin Jews, and others) are indicated wherever they are known.

Many popular myths are dispelled in the book, including the false ideas that all Ashkenazic maternal lines came from Italian converts to Judaism, that the Slavic DNA in Ashkenazim came from Cossack men during pogroms, and that the Ashkenazic people descend substantially from Turkic Khazar converts to Judaism.

Data concerning the maternal lineages of the Krymchak Jews are also published here for the first time.

Acknowledgments

This book uses data from published studies as well as from public and restricted-access databases including Family Tree DNA, mtDNA Community, 23andMe, GEDmatch, YFull's MTree and YTree, *YSEQ*, AmtDB, the Human Genome Diversity Project, and the National Center for Biotechnology Information's GenBank and *BioProject*. I thank all of the people who have created, built, and maintained those excellent databases, among whom are Bennett C. Greenspan, Doron M. Behar, Vadim Urasin, and Edvard Ehler. Greenspan and I have communicated about Jewish DNA, including haplogroups, many times since the late 1990s and I have always appreciated his information and opinions.

Additional data and reading recommendations came from the following people with whom I corresponded: Sandy Aaronson, Yoav Aran, William Miles Boyce, Corey Steven Bregman, Leo Raphael Cooper (a co-administrator of several Family Tree DNA projects including "Jewish Ukraine-West" and "E-Y14891"), Janet Danner, the late Martin Davis, Igor Dorfman, Stuart M. Drucker, Eugene Dubrovitzky, Arthur Fagen, Marc Friedman, Joshua Glasser, Samuel T. L. Hilsenrath, James M. Hornell, Debra Katz, John Dale Kessel, Arye Kranz, Leon Kull (co-administrator of Family Tree DNA's "Karaites of Eastern Europe" project), Joshua Lipson (a co-administrator of several Family Tree DNA projects including "Jewish Ukraine-West" and "E-Y6923 Jewish Haplogroup"), Ian Logan, Ariel Lomes, my cousin Paul Avrell Morrison, Barry L. Myones (a co-administrator of Family Tree DNA's "Romaniote DNA Project"), Oshri Pesach Naparstek, Israel Pickholtz, Judd R. Rothstein, Gary Spitalnic, David B. Sterlin, Jonah Aaron Stern, Emelie Wallgren, Joshua Gerson Weinstein, Jeffrey D. Wexler (manager of the "Levite DNA" research project at LeviteDNA.org), Melissa Yagusic-Cupiccia, Jacob E. Wilamowski, and Irwin Zweigbaum. They helped to enrich my knowledge of dozens of Ashkenazic haplogroups. Cooper, Lipson, Stern, and Wexler also read drafts of my manuscript and provided feedback on them. Anthony Lizza (administrator of Family Tree DNA's "Sicily" project) provided some of my information on Sicilian haplogroups.

Data that these sources provided have been anonymized in this narrative so as not to identify particular living DNA testers.

Thanks to the five anonymous scholars from the fields of history and genetics who made suggestions to improve the manuscript during its peer review process. Also thanks to Alessandra Anzani, my editor at Academic Studies Press, and Ilya Nikolaev, my production editor.

Special thanks to Phillip Arieff, David Chambers, Jared Hirsh, Howard Sharpe, Jonah Stern, and my cousins Michael Balk, Suzanne Fefer, Lawrence Gordon, and Adam Kerchner for contributing portions of the subvention.

KAB, April 2022

Chapter 1: An Introduction to Ashkenazic History and Genetics

Early settlements, migrations, and subgroups

The Ashkenazic ethnicity formed around the ninth century in Central Europe, particularly in Germany, and later spread into other lands, including Hungary, Czechia, Poland, Lithuania, and Russia. Many of the ancestors of Central European Jews had lived in Italy and France.

The earliest reference to Jews in lands that became part of Germany dates from 321: an edict issued that year by the Roman emperor Constantine I refers to Jews living in the city of Cologne (Köln) in western Germany.[1] Scholars have expressed different views as to whether or not the pioneering Jews of Cologne were direct ancestors of later Jews in that city or whether they fled to Gaul.[2] Most other Jewish communities in Germany are attested only much later. Records show that Jews were living in Mainz by 906, in Worms by 960, in Magdeburg by 965, in Regensburg by 981, in Speyer by 1084, and in Frankfurt am Main by 1241. Many early German Jews worked as merchants and chose to live in cities like these because they were connected to major trade routes by river and by land.[3] For example, Mainz, Worms, and Speyer were adjacent to the Rhine River, Magdeburg was by the Elbe River, and Regensburg was by the Danube.

German Jews developed the first version of the Yiddish language largely from Middle High German mixed with some Hebrew and other elements and wrote it using the Hebrew alphabet.

Jews were living in the Czech city of Prague by 970. Jews in Czechia learned the Old Czech language and developed the West Knaanic culture.[4]

Jews lived in Magyar-ruled Hungary by the 950s.[5] The first Jews to live in Poland's capital city of Kraków moved there in the twelfth century and Jews also settled in other Polish cities and towns during the twelfth-fifteenth centuries, such as in Płock by 1237 and Warsaw by 1414.[6] Jewish communities also sprang up in multiple cities in Kievan Rus' that later became part of Ukraine, including in Kyiv by 962 and in L'viv by the 1350s.[7] Many Jews in Kievan Rus' learned how to speak East Slavic dialects and developed the East Knaanic culture, which lasted for a particularly long time in some parts of Belarus.[8]

Jews who belonged to the West Knaanic and East Knaanic cultures assimilated many Slavic personal names and some of these names later became Yiddishized.[9] Thus, some Ashkenazic women who spoke Yiddish had names of Old Czech origin like Zlate and Prive and others had names of East Slavic origin like Badane and Vikhne, but others had names of German origin like Golde and Freyde, names of Romance (Old Italian and Old French) origin like Yentl and Toltse, and names of Hebrew origin like Sure and Rukhl.[10] As I will show, this mixture of names coincidentally resembles the ethnic mixture of most Ashkenazic maternal lineages.

After 1492, Ashkenazic Jews intermarried with Sephardic Jews from Spain and Portugal who had moved into Germany, Poland, and Ukraine.[11] Some descendants of these Sephardim later moved into Belarus, Lithuania, and Latvia.[12]

Nearly all Jews in the Ashkenazic cultural sphere eventually adopted the Yiddish language, giving up French, Judeo-Spanish, West Slavic, and East Slavic, except for some traces of those languages that remained in Yiddish and in Ashkenazic given names. They also all adopted the Ashkenazic rite of rabbinical Judaism, which has minor differences from the Sephardic rite, and Ashkenazic-specific *takkanot* (enactments) and *kheremim* (bans) that were decided upon by Ashkenazic rabbis.

German Jews from families with long continuous histories in Germany and some Jews in neighboring countries, including Switzerland, are known as Yekkes. The Ashkenazic Jews living further east developed new subcultures as well as new dialects of the Yiddish language. Oberlanders were the Ashkenazim living in territories that are today in western Slovakia and far-eastern Austria's Burgenland region. Unterlanders were the Ashkenazim living in eastern Slovakia, northwestern Transylvania, and southwestern Ukraine's Carpathian Ruthenia region. Galitzianers were the Jews of the Austrian Empire's crown land of Galicia in southeastern Poland and western Ukraine. Litvaks were the Jews of Lithuania, Latvia, Estonia, Belarus, and certain adjacent areas of Poland, Ukraine, and Russia, and their dialect of Yiddish was called Litvish. Two additional Ashkenazic subcultures that were much smaller emerged: the Frisian Jews of Friesland in the northern Netherlands and the East Frisian Jews of rural northwestern Germany, both of which had their own distinctive dialects of Yiddish. It is important to emphasize here that all of these populations substantially descended from medieval German Jews.

Over the centuries, including during the 1400s and 1500s, some Ashkenazic men resettled in non-Ashkenazic Jewish communities, such as those in Turkey, Syria, Tunisia, Algeria, and the Crimea. Their descendants assimilated into the local Jewish cultures but sometimes carried surnames that identified their Ashkenazic paternal origins, such as Eskenazi among the Jews in Aleppo, Syria and Esquinazy among the Jews in Oran, Algeria. Some Ashkenazim returned to southern Europe, including German Jews who moved into northern Italy beginning in the late 1200s[13] and Ashkenazim who lived in Crete in the early 1500s.[14]

An overview of Ashkenazic autosomal DNA and Y-DNA

Approximately half, or a little more than half, of the genetic ancestry of Ashkenazic Jews from Eastern Europe traces back to the ancient Middle East. Ashkenazim have partial similarities to the autosomal DNA in ancient bones from Tel Megiddo (northern Israel)[15] and other parts of the Middle East. This type of ancestry is represented by a large proportion of Ashkenazic Y-chromosomal haplogroups, passed down by males, as well as by a minority of their mitochondrial DNA (mtDNA) haplogroups, passed down by females. Many of these haplogroups stem from Canaanite, Edomite, Arabian, Mitanni, Iranian, and Egyptian progenitors who contributed to the gene pool of Jews in ancient Israel and its southern successor kingdom, Judea. Modern matches to these haplogroups include Arab-speaking non-Jews of the Levant such as Syrians, Lebanese, and Palestinian Arabs but often Gulf Arabs as well.

At the same time, comprehensive autosomal studies show that most of the other half of Ashkenazic ancestry derives from European sources, particularly from southern Europeans (including Italians and Greeks).[16] The proportion of genetic contribution from northwestern Europeans has been estimated to be between 3 and 8 percent, while the contribution from eastern Europeans has been estimated to represent between 8 and 12 percent, on average.[17] However, some Ashkenazic individuals have inherited only 5 percent or less of Eastern European DNA, as we know from studying many kits using autosomal admixture calculators. German Jews and Dutch Jews have the least of it. These elements from northern, central, and eastern Europe, along with traces of East Asian ancestry, are factors that distinguish the genetic profile of Ashkenazic Jews from those of Sephardic Jews, Italki Jews, and Romaniote Jews.

Ashkenazic autosomal DNA can be modeled as about 87.2 percent related to Sephardic autosomal DNA in terms of sharing the same ancestral sources from approximately the start of the common era through the sixth century, and Ashkenazim are about 79.6 percent related to Romaniote Jews from Greece.[18] Italkim, Sephardim, and Ashkenazim share Northern Italian autosomal DNA, which is lacking in Romaniotes. According to calculations performed by Ariel Lomes, this Northern Italian ancestry amounts to about 9 percent of Sephardic autosomal DNA, 10.8 percent of Ashkenazic autosomal DNA, and 13 percent of Italki autosomal DNA.[19]

The careful study of Y-chromosomal and mitochondrial DNA haplogroups helps us to more fully understand who the ancestors of the Ashkenazic people were before they became Ashkenazic.

R1a-Y2619 is the most frequent Y-chromosomal haplogroup family in Ashkenazic men as a whole (7 to 9 percent) and is found in around 52 to 65 percent of Ashkenazic Levites, including those with documented lineages from fifteenth-century Levite men from Germany and Czechia. The majority of them belong to the subclades R1a-Y2630 and R1a-FGC18222. R1a-Y2619 originated in the ancient Middle East.[20] Brother subclades of R1a-Y2619 are found in modern Persians from Iran's Kerman Province, Azeris from Iran's West Azerbaijan Province, and Yazidis[21] from Iraq's Kurdistan region, and also in Spaniards and people in Armenia and eastern Turkey's Van province. The common ancestor of the Jewish and non-Jewish subclades was the haplogroup R1a-CTS6 (also called R1a-M582) and probably lived in either western Persia or in

Mesopotamia around 950 B.C.E., as estimated by the company YFull's YTree.[22]

J1-Y5400 is the most common Ashkenazic Cohen Y-DNA haplogroup family. Its frequency in the total Ashkenazic population is as high as 6.2 percent. All Ashkenazic members belong either to the branch J1-Y5399 or to the branch J1-Y31161. J1-Y5400 is a branch of J1-M267's descendant J1-P58's expanded Cohen Modal Haplogroup subclade J1-Z18271 and is definitely of Middle Eastern origin.[23] Some branches of J1-Z18271 are found in Sephardic Jews and Portuguese and Hispanic Catholics who descend from Sephardim. All indications are that these Ashkenazic and Sephardic haplogroups descended from the ancient Israelites and would have been found in Diaspora Jews in the Roman Empire.

Another especially prevalent Y-DNA haplogroup in Ashkenazim is E-Y6938, within the broad E1b family of haplogroups. Some brother subclades of E-Y6938 are found in North African Jews (including Libyan Jews) and Sephardic Jews from Greece and multiple populations that descend from Jews who converted to Catholicism, including Sicilians and Latin American Hispanics. The most recent common ancestor of E-Y6938 and these other haplogroups carried E-Y6923 and was probably a Jew who lived in Mediterranean Europe in late antiquity, around the year 350. Before then, we know that the patrilineal ancestors lived in the Middle East because they share E-Y6926 as an ancestor with men from the United Arab Emirates and Oman who have patrilines of Baloch origin.

Another Y-DNA haplogroup that is frequently found in Ashkenazim is Q-Y2200. They belong to branches under its subclades Q-YP1003 and Q-Y2198. A different branch of Q-Y2200, Q-BZ72, is found in Sephardic Jews from Greece as well as some populations descended from Jews who became Catholics, including Sicilians, Mexicans, Portuguese, and Brazilians. Q-Y2200's parent haplogroup, Q-Y2225, is found in Sicilians and Lebanese Christians. Indications are that the patrilineal ancestors of Q-Y2225 lived in the Middle East[24] and that Ashkenazim inherited this haplogroup from the Israelites. It is significant that Q-Y2225's parent haplogroup, Q-Y2209, is found in Iraqi Arabs, Emiratis, Omanis, Jordanians, and Lebanese.

The important Ashkenazic Y-DNA haplogroup G-Y12975 is also of Middle Eastern origin. It is distantly related to haplogroups that are found in such peoples as Syrians, Lebanese, Armenians, Sicilians, Pashtuns, and Tajiks.

J2-L556 is another prominent Ashkenazic Y-DNA haplogroup that is likely of Middle Eastern origin. Its branches are called J-Y9005 and J-Y11782. A majority of Ashkenazic carriers belong to subclades of the former and a minority to those of the latter. J-Y9005's subclade J-A16313 has been found in men from Italy and Bulgaria who descend from a medieval Ashkenazic rabbi who lived in northern Italy. J2-L556 is distantly related to haplogroups that are found in Italians (including Sicilians), Saudis, Iraqis, and Assyrians.

The common Ashkenazic Y-DNA haplogroup J1-L816 is probably from the Middle East, too. Most Ashkenazim within this haplogroup family belong to five branches under its subclade J-ZS2728 but a particular German Jew was found to carry its other subclade, J-Y34527, which also has a Sephardic-origin variety found in Mexican Catholics from Nuevo León and in New Mexican Hispanos. A sixth branch of J-ZS2728 is found in Romaniote Jews from Greece. J1-L816 shares an ancient ancestor with a haplogroup that is found in Kurds from eastern Turkey and even more ancient ancestors with Saudis, Tunisian Jews, Turkish Jews, and Italians.

Another common Ashkenazic Y-DNA haplogroup that appears to originate from the Middle East is E-Y14891, which splits into the subclades E-Y16781 and E-Z36149. E-Y16781 has also been found in a Jewish patriline from Venice, Italy. E-Y14899, the parent haplogroup of E-Y14891, has other branches that are found in Palestinian Arab Christians, Greek Cypriots, Kuwaitis, Emiratis, and Arabs from Bahrain.

My Y-DNA haplogroup appears to be E-Y87732, a branch of E1b-Z830 that is downstream from E-Y35934, because my closest Y-DNA match inside Family Tree DNA is definitely in that branch according to his "Big Y" test, as are several more of my close paternal-line matches. It is a clear case of Ashkenazic descent from Israelite men. Its distant non-Jewish matches are mainly Middle Easterners, mainly Arabs from Saudi Arabia, Syria, Kuwait, and Iraq but also a few men from Armenia and one each from Iran and India.

The aforementioned haplogroups E-Y14891, G-Y12975, J2-L556, and Q-Y2200 plus another one called J2-Y15223 have carriers in modern

German Jewish men with documented Rhineland patrilineal genealogies, according to research by Joshua Lipson in collaboration with Leo Cooper and Anthony Lizza, leading them to the conclusion that they were likely found in early Rhineland Jews. R1a-Y2619 and J1-Y5400 were probably also present among the early Rhineland Jews. By contrast, E-Y6938 is very rare among Rhineland Jewish families despite it being the second-largest lineage among Eastern Ashkenazim. Being the only major lineage to have this discrepancy, this may suggest that it would have been found instead among early Jews in Bavaria, Austria, or Czechia, and this is Lipson's team's tentative conclusion.

There are other Ashkenazic Y-DNA lineages of Middle Eastern origin beyond the ones I mentioned but also some lineages that have origins outside of the Middle East. A particularly exceptional one is G2a-FGC1093, which is found in Litvak Jews from Belarus and Lithuania as well as in North Ossetians and Kumyks from the North Caucasus region. YFull's YTree estimates that their common ancestor lived around the year 700. It now appears that this common ancestor would have been an Alan man. According to the *Schechter Letter*, which was authored by a Khazarian Jew, some Alans living in the kingdom of Alania converted to Judaism around the 930s under the influence of the Khazars.[25] G2a-FGC1093's parent haplogroup, G-FGC1144, is likewise found in Ossetians. The parent of G-FGC1144, G-FGC1159, is found in Svans from Georgia and somebody from Azerbaijan and has a branch called G-BY96310 that is found among two more ethnicities in the Caucasus: Adygei (western Circassians) and Karachays. A predominantly Caucasian distribution continues with more distant matches including more Svans, more Adygei, and more Ossetians plus some Chechens, Ingushetians, Abazinians, Georgians, and an Ossetian-descended Jassic patriline from Hungary, as well as men from unlisted ethnicities from Trabzon Province in northeastern Turkey, where some Circassians live. The aforementioned haplogroups are branches of G2. At least one variety of G2 was found among medieval Alan men[26] but their specific subclade(s) have not yet been published as of the time of this writing. Leo Cooper constructed statistically well-fitting genetic autosomal DNA models according to which Ashkenazim from Lithuania and Belarus descend 4.2 percent from medieval Alans, dropping to 0.8 percent in Ashkenazim from Poland and Ukraine and 0 percent in Ashkenazim from Germany.[27]

The Ashkenazic Y-DNA haplogroup J2-Y33795 most likely derives from a convert to Judaism in Roman-era Italy. Although members of this haplogroup identify as Cohenim, there is no evidence to connect it to ancient Israelite priests. The most recent common ancestor of J2-Y33795 is estimated to have lived around 250 C.E. according to YFull's YTree and it is absent from Middle Eastern non-Jews. A branch of J2-Y33795 that is called J2-Y37837 was found in an Italian who tested with YFull. J2-Y33795's brother subclade J2-Y34371 is present in Switzerland. The more distantly related subclades J2-Y22038 and J2-Y22881 are found in Italians (Northern Italians and Central Italians, respectively). Another related subclade, J2-Y45181, was found in an Etruscan from a necropolis near Civitavecchia, Rome who died around 600 B.C.E.[28]

Genetic studies have demonstrated that the early Jewish populations of southern Europe were founded by Jewish men of predominantly Israelite origin who sometimes married non-Jewish women who converted to Judaism, usually voluntarily. A few non-Jewish men also converted to Judaism and joined these communities. These studies also show that intermarriage occasionally continued after proto-Ashkenazic Jews migrated north of Italy, even though this phenomenon has sparse documentation.

Males were not responsible for the majority of the northern and eastern European ancestral introgressions into the Ashkenazic gene pool. For one thing, there is not a large amount of Y-chromosomal diversity among the Ashkenazic paternal haplogroups that originate from those regions, indicating a low number of individual male contributors from non-Jewish populations of Germany, Poland, and environs. According to research conducted by Leo Cooper, there are fewer than ten such lineages and they collectively are found in only between 4 and 8 percent of Ashkenazic men. There could not have been many intermarriages or extramarital affairs or (already less likely) rapes between non-Jewish men and Jewish women that resulted in offspring. This also explains why the majority of Ashkenazic Y-chromosomal lineages trace to Mediterranean and Middle Eastern regions.

It is a widely believed myth that Ashkenazim with Slavic DNA, blue eyes, or blond hair inherited those traits as the result of Cossack rapes of Jewish women in the 1640s. However, there is no evidence for any Y-DNA or autosomal DNA inputs from Cossacks.[29] The closest matching population for Ashkenazim's Slavic autosomal DNA are the Poles, most of this DNA came from women rather than men, and only the women lived near the time of the Cossacks.

David Wesolowski modeled Ashkenazic autosomal admixture using a large data set including from populations collected by the Estonian Biocentre Human Genome Diversity Panel and found the following elements with a distance of 0.2874 percent: 33.6 percent Samaritan-like, 30.45 percent like Italians from Tuscany, 15.65 percent like Arabs from Israel, 11.75 percent Polish, 7.9 percent like Chalcolithic-era Anatolians, 0.6 percent like Avars from the North Caucasus, and 0.05 percent like Turkic-speaking Bashkirs but zero percent for Cossack and Uyghur affinities.[30]

Some Eastern Ashkenazic men, including Latvian Jews, and some Western Ashkenazic men, including German Jews with surnames indicating deep roots in Germany and a German Jew with a surname (Prager) indicating roots in Prague, belong to varieties of the paternal haplogroup R1a-M12402. The most recent common ancestor of this haplogroup is estimated to have lived around the year 650 according to YFull's YTree. Additional data from Family Tree DNA suggest that the first Ashkenazi to carry it may have lived in around 900. Its source was a Slavic man but not necessarily a Pole. It is certainly not Cossack. Joshua Lipson's research team believes it is more likely that this man joined the West Knaanic Jewish community in Bohemia as opposed to the East Knaanic Jewish community in Ukraine. The surname Prager confirms the connection to Bohemian Jews. R1a-M12402's parent haplogroup, R1a-YP4848, is found in non-Jewish Ukrainian, Southern Russian, Lithuanian, and Bulgarian men. The YTree estimates that the most recent common ancestor of R1a-YP4848 carriers lived in about 300 C.E. More distant haplogroup relatives of R1a-YP4848 are found in ethnic Bulgarians, Russians, Belarusians, Poles, Finns, Volga Tatars, and Lithuanian Tatars and people from Ukraine, Slovakia, and Serbia, among others.

R1b-FGC8564, a branch of R1b-Y5051, is an example of an Ashkenazic paternal haplogroup of ultimately northwestern European origin. Ukrainian Jews and Turkish Jews share the branch of it that is called R1b-Y99541. Lithuanian Jews, Belarusian Jews, and Polish Jews belong to other branches of it. Although the closest non-Ashkenazic and non-Jewish matches to R1b-FGC8564 are Portuguese, Brazilian, Spanish, and Mexican, indicating an Iberian connection, more distant matches include Germanic men: Swedes, Germans from southwestern and northwestern Germany, and Englishmen, plus Norwegians and Danes as even more distant matches followed by additional Swedes and Germans. According to YFull's YTree's estimates, one of these Englishmen shared a common ancestor with the Ashkenazim who lived around the year 400 while other Englishmen shared a common ancestor with the Ashkenazim who lived around 100 C.E. and still other Englishmen share a common ancestor with them who lived around 550 B.C.E. The Hispanic and Lusophonic matches probably had a Visigothic ancestor.

Conversions to Judaism and intermarriages between Jewish men and non-Israelite women in Western Jewish communities

According to rabbinical law, membership in the Jewish religion is transmitted through the female line, while tribal status (*Kohen*, *Levi*, or *Yisrael*) is transmitted through the male line. Converts to Judaism become full Jews in most respects in the eyes of rabbinical law.

Specific male and female converts to Judaism in the Roman Empire's European and African territories are known. A woman named Beturia Paulina who lived in the city of Rome converted at the age of 70, around the year 50 C.E., and took the new name Sara.[31] Another woman in Rome, Felicitas, converted when she was 41 and adopted the name Peregrina.[32] A female proselyte named Crysis also lived in Rome.[33] In the first century C.E., in the city of Cyrene in what is now northeastern Libya, a female proselyte named Sarra died at the age of 18.[34]

During the early centuries of Jewish habitation of Germany, when relations between Jews and Christians were better than they later became, and when some Germans still practiced paganism, there were occasional male and female German converts to Judaism. The Rhineland massacres of 1096 were a turning point. In 1096, many Jews were killed by Christian crusaders and local German Christians for refusing to either convert to Christianity or return to Christianity. The Jewish community of Köln recorded in its *Memorbuch* ("Book of Remembrance") that two converts to Judaism, including a woman named Hatziva, lost their lives at this time.[35]

When Jews were expelled in December 1288 from the counties of Anjou and Maine in northwestern France by Count Charles II, the edict of expulsion claimed that Jewish men had been having intimate relationships with the region's Christian women.[36]

During many eras, the Polish government (and its post-partition successors) along with the Jewish and Catholic religious authorities prohibited cross-religious intermarriage. The government and church also prohibited Christians from becoming Jews. Rabbi Moses Isserles of Kraków (1530–1572) observed that "in these lands . . . it is forbidden to convert non-Jews," while Rabbi Solomon Luria of Lublin (1510–1573) warned Jews against the potential dire consequences of accepting converts or promoting proselytism.[37] However, during the Protestant Reformation period, especially the sixteenth century, the Roman Catholic Church temporarily lost its grip over some of the Polish people. This timeframe corresponds to autosomal DNA analysis of identical-by-descent segment lengths which indicates that Poles contributed DNA of Slavic and central European origin to the Ashkenazic people around the fourteenth-seventeenth centuries.

Documents concerning intimate relationships between Jews and former or current Christians during this period are infrequent and pervaded with negative sentiments. In 1548, the bishop of Vilnius wrote a letter in which he complained that Catholic women were often marrying Jews and other non-Catholics and raising their children to believe in their fathers' religions rather than Catholicism.[38] Within Poland and Polish-ruled western Ukraine, during the sixteenth-eighteenth centuries, there were instances of Slavic (including Polish) women, some of whom were housemaids and tavern servants working for Jews, having extramarital affairs and marriages with Jews, sometimes after the women converted to Judaism.[39] Contemporary bishops in Przemyśl and Chełm and a priest in Sandomierz wrote complaints against the sexual relations that were occurring between Catholic women and Jewish men.[40] Many of the women, Maryna Wojciechówna and Paraska Daniłowna among them, were caught by the authorities and put to death before leaving any descendants, but some of the other women had children.

Total endogamy for Ashkenazic populations in central and eastern Europe set in during the late eighteenth and nineteenth centuries when there were no further additions of non-Ashkenazic ancestry (European, Sephardic, Italki, Mizrahi).

In the following chapter, I will explore the maternal lineages of Ashkenazim and what they reveal about Ashkenazic origins. Most of them originate with Middle Eastern and European populations, but several of them point to unexpected origins like China and North Africa.

Chapter 2: Encyclopedia of Ashkenazic Maternal Lineages

A deep dive into every surviving variety of Ashkenazic mtDNA that could be verified

Human mitochondrial haplogroups have been sequenced by scientists since 1981. In early years of research into the mtDNA haplogroups that are found in the Ashkenazic population, many scientists designated some of their haplogroups merely by a single letter, such as K or M, or with just one letter and one number, such as H1 or M1. This lack of detail did not enable many specific conclusions to be made. Over time, new branches of the haplogroups were discovered and named. Geneticists' branching of mtDNA haplogroup subclades has significantly advanced since the time of early studies of Ashkenazic DNA by such authors as Behar,[41] Coffman-Levy,[42] and Costa.[43] This analysis can be used to accurately trace individual lines of genetic evidence that European women came into the Jewish fold when marrying Jewish men, whether clandestinely or openly, even when vital records and historical accounts are lacking. It can also show that certain other Ashkenazic lineages had a pathway from ancient Middle Eastern Jewish communities, including but not limited to those of the Kingdom of Judah in the land of Israel.

Most Ashkenazic mtDNA haplogroups have either not been examined or not been fully parsed out in previous published research. This chapter incorporates the latest available knowledge of their origins and their geographic and ethnic distributions both in ancient and modern times accompanied by analysis of phylogenetic trees, mutations, genetic distances,[44] and dating. It is likely that future DNA samples will further elucidate the origins of certain haplogroups. Additional sampling of undersampled populations could very well alter some of the conclusions.

In an entry in this chapter, I will present new evidence for the North German origin of the Ashkenazic maternal haplogroup H7j. I will also reveal my discovery that the maternal haplogroup H2a1e1a, also probably of German origin, exists in some Ashkenazim. I will discuss multiple Ashkenazic haplogroups that originated with Polish women. I will introduce my discovery of the Ashkenazic maternal haplogroup U5a1d2b, which probably came from eastern Europe. I will also summarize recently obtained ancient genetic samples that show the early geographic distribution of other Ashkenazic haplogroups in Europe, North Africa, and the Middle East. Furthermore, I will explain which of the haplogroups are shared with other types of Jews, including with Sephardic Jews and Mizrahi Jews. I present my news that the Ashkenazim with the maternal haplogroup H1u2 match Crimean Karaites.

There are numerous Eastern Ashkenazic mtDNA lineages that are not found among German Jews. This is because Eastern Ashkenazim are more genetically diverse than Western Ashkenazim.

This study is exclusively about the mtDNA lineages that already existed in the Ashkenazic population in central and eastern Europe in the early nineteenth century before any intermarriages had taken place between individual Ashkenazic men and Sabbatarians of Székely origin, Subbotniki of ethnic Russian origin, and other recent converts to Judaism from the second half of the nineteenth century onward. Inclusion of any mtDNA haplogroup in this encyclopedia was dependent on viewing the admixture results of fully sequenced Ashkenazic carriers of that haplogroup in reliable calculators and wherever possible combining that with genealogical documentation. The phrase "full Ashkenazi" refers to individuals with genetic and genealogical indications of having sixteen Ashkenazic great-great-grandparents, who most often score 99–100 percent Ashkenazi Jewish at Family Tree DNA, 23andMe, and AncestryDNA. All of these Ashkenazim were genetically tested during the first three decades of the twenty-first century.

Statistical frequencies of mtDNA haplogroups in the Ashkenazic population

From Wim Penninx's large data set of Family Tree DNA customers with high amounts of Ashkenazic ancestry,[45] I found 3,486 samples with fully sequenced haplogroups that exist in Ashkenazic matrilines after I filtered out partially sequenced haplogroups and irrelevant haplogroups. I then calculated each haplogroup's percentage. In order of descending frequencies, there are 723 listed K1a1b1a carriers (20.7%), 214 K1a9 (6.1%), 210 N1b1b1 (6%), 185 K2a2a1 (5.3%), 139 HV1b2 (3.99%), 84 V7a (2.4%), 80 J1c7a (2.3%), 77 L2a1l2a and L2a1l2a1 (2.2%), 67 H7e (1.9%), 66 HV5a (1.89%), 57 H1e4 combined with H1e4a (1.64%), 54 U7a/U7a5 (1.55%), 52 H1as2 (1.49%), 50 J1c14 (1.43%), 50 H1aj1 combined with H1aj1a (1.43%), 47 H3w (1.35%), 46 H3p (1.32%), 45 R0a4 (1.29%), 44 I1c1a (1.26%), 43 J1b1a1 (1.23%), 43 R0a2m (1.23%), 42 H1b2a and H1b2a1 (1.2%), 41 H40b (1.18%), 35 T2b25 (1%), 34 H6a1a5 (0.98%), 33 H5c2 (0.95%), 33 H7c2 (0.95%), 33 W3b1 (0.95%), 32 J1c1 (0.92%), 31 H26c (0.89%), 30 W3a1a1 (0.86%), 28 H7 with mutations that define H7j and H7j1 (0.8%), 28 U1b1 (0.8%), 25 H11b1 (0.72%), 25 X2b7 (0.72%), 21 V1a1 (0.6%), and the following have frequencies below 0.6%: 20 H6a1a1a, 20 H10a1b, 18 H41a, 18 H5a7, 18 N9a3, 18 T2e1b and T2e1b1, 18 T1b3, 15 U3a1a, 15 H65a, 15 T1a1j, 15 T1a1k1, 14 V7b, 14 W1h, 13 K2a, 13 H1ai1, 13 H1bo, 13 U5a1f1a, 12 T1a1, 12 H56, 12 M33c, 12 U6a7a1b, 12 J1c13, 12 H25, 11 H1b1-T16362C/H1b1a, 11 M1a1b1c, 11 T2a1b, 10 H11a2a2, 10 HV1a'b'c, 10 U2e1a1, 9 H1b, 8 H3, 8 H6a1b3, 8 H11a1, 8 J2b1e, 7 H5-C16192T, 7 H6a1a3, 7 H15b, 7 J1c-C16261T, 7 K1a4a, 7 U4a3a, 6 H4a1a3a, 6 H5-T16311C!, 6 U5a1b1, 6 V15, 5 H5a1, 5 HV0-T195C!, 5 J1c3e2, 5 T2a1, 5 T2b16, 5 U3a1, 5 X2e2a, 4 H1u2, 4 H2a2b1, 4 H6a1b2, 4 HV5, 4 T1a, 4 T1b, 4 T2b4, 4 U5a2b2a, 3 H1f, 3 H1bw, 3 H4a1a1a, 3 H5, 3 H13a1a1, 3 H47, 3 N1b1a2, 3 T1a1b, 3 T2g1a, 2 A-T152C!-T16189C!/A12'23, 2 H1ax, 2 H3ap, 2 N1b2, 2 T2b4a, 2 U5a1a2a, 2 U5a1b, 2 U5a1b1c2, 2 U5b1b1-T16192C!, 2 U5b1e1, 2 U8b1b1, 2 V18a, 1 H1bd, 1 H4a1a1, 1 H13a1a1a, 1 I5a1b, 1 J1c4, 1 J1c5, 1 K1a1b1, 1 T2b3-C151T, 1 U5b2a1a, 0 H2a1e1a, and 0 U5a1d2b. All of these haplogroups are discussed separately below and my data set incorporated an increased number of samples with verifiable Ashkenazic matrilines.

A12'23

Haplogroup A is one of the rarest[46] and most mysterious Eastern Ashkenazic maternal lineages. Some Ashkenazic A carriers lived in Poland, the Russian Empire's Pale of Settlement, and the city of Saint Petersburg. Indications are that A would not have been found among the medieval German Jews. The Ashkenazic variety's complete sequence's current technical name in Family Tree DNA is A-T152C!-T16189C! (the exclamation points indicate reversed mutations, known as back mutations). Its former designation was A4 and its current alternate designations include A12'23 and A12/23 (the latter was used by Kristiina Tambets' research team) and A-a1b (used by YFull's MTree).[47]

It is one of three Ashkenazic matrilines of East Asian origin. The largest number of carriers of A live in East Asia, and that is also where its greatest degree of variety is found. Some of the East Asian and Southeast Asian populations that carry varieties of A include Mongolians, Japanese, Ryukyuans, Koreans, Han Chinese, Tibetans, Nakhi of China, Yi, Hmong, and Vietnamese. Branches of A are also present among some Siberian populations such as Evenks, Buryats, and Mansi and among Volga Tatars, Anatolian Turks, eastern Persians, and Tajiks. A2 is a famous branch of A that is present among Inuit and Amerindian tribes and Spanish-speaking mestizos who descend in part from Amerindians.

Carriers of A-T152C!-T16189C! who participate in Family Tree DNA's mtDNA Haplotree[48] list their matrilineal ancestors' home countries as Poland, Romania, Hungary, Austria, Russia, and Uzbekistan. I assume that all of them are Ashkenazim except for the carrier with roots in Uzbekistan. GenBank includes a couple of non-Jewish types of carriers of A-T152C!-T16189C! and the relevant mutation T16189C! is confirmed to be present in them. Two are Pamir highlanders from Central Asia.[49] Another is a member of the *Khon Mueang* (Northern Thai) people of Thailand.[50] The Northern Thai are related to the Dai people of Yunnan Province in southwestern China and results with good fits are sometimes obtained when Dai people are included as a minority component to model the admixture of Ashkenazim as an alternative to a Han component. Numerous Udmurt and Komi carriers of this haplogroup were sampled by Tambets' team.[51] Those are ethnic groups from Russia.

The Haplotree and MTree both reveal descendant subclades of A-T152C!-T16189C! to include haplogroups A12 and A23. Family Tree DNA's carriers of A12 have matrilines from Germany and Czechia and their carriers of A12's subclade A12a have matrilines from England and Ireland. A12a is found among the Nenets, Mansi, Selkup, and Yakut peoples of Russia and the Kyrgyz of Kyrgyzstan. MTree found the newly identified branch A12a1 in modern Russia among the Evenk people and A12a2 in ancient Russia and modern Hungary. The subclade A12b is found in Buryats in Russia and was in medieval Hungary. As of the time of this writing, the only A23-carrying participant at Family Tree DNA is someone who lists Russia for their matrilineal origin. Tambets' team found A23 in a Ket person from Siberia. MTree calls a branch of it A23a and found it in a Buryat person from Russia and a Qashqai person from Iran. MTree shows A23 as a daughter subclade of A-a1b3. A-a1b3, also called A-a1b3* and defined by the mutation A3213G that originated as early as 16,850 B.C.E., was found in ancient Kazakhstan, ancient Xinjiang, and ancient Siberia. Geoffrey Sea wrote that his matriline is Hungarian Jewish and that he was assigned to A-a1b3*.[52] Subclade A-a1b3a, defined by the extra mutation T14025C, formed around 10,250 B.C.E. and was found in an ethnic Turkmen from Samarkand, Uzbekistan who tested with YFull.

A Buryat person from China's Inner Mongolia Autonomous Region who was tested by Derenko's team carries both the T152C! and T16189C! mutations that Ashkenazim share but also has a unique mutation called T2563C.[53] MTree therefore places this Buryat into a new daughter subclade of A-a1b that it calls A-a1b2. MTree also invented the name of haplogroup A-a1b1, defined by the mutation A15064G, and assigned it to multiple Uyghurs from Xinjiang.

MTree reveals that Buryat people from Russia carry A-a1*,[54] the parent haplogroup of the Ashkenazic haplogroup. A scientific team found A-a1* in multiple Chinese people[55] and MTree found it in a person from the Hebei province in northeastern China.

If the assignment Sea received is accurate, the first Jewish A was probably a Khazarian woman.

H1ai1

H1ai1 is not that common among Ashkenazim, nor in other populations. It was not detected to be carried by any of the Ashkenazic participants in Costa's team's mtDNA research[56] but inside Family Tree DNA I found confirmation that some full Ashkenazim do carry it, including at least one with an Ashkenazic matriline from Lithuania and at least one with an Ashkenazic matriline from Poland. At Family Tree DNA, H1ai1 is also found in small numbers of non-Jewish customers who trace their matrilines to Ireland, Wales, and England. Leo Cooper showed me that one of the Ashkenazim with this haplogroup who tested with Family Tree DNA matches some of these non-Jewish carriers of the same haplogroup in the Full Coding Region screen, including at least two with matrilines from Ireland at a genetic distance of 1 and a matriline from Wales at a genetic distance of 3.[57]

Its parent haplogroup, H1ai, has been found in non-Jewish matrilines from Ireland, Scotland, England, Germany, and Spain. It has not been found in matrilines from outside of Europe.

A male Viking skeleton from the tenth century who was found at the Buckquoy site in Birsay, Orkney, northeastern Scotland was confirmed to belong to H1ai1.[58]

The first Ashkenazic H1ai1 was presumably a northwestern European convert to Judaism.

H1aj1

Some full Ashkenazim who had their complete mitochondrial sequences tested by Family Tree DNA were determined to belong to H1aj1, which originated about 8,000 years ago according to YFull's MTree. Some of them trace their family trees matrilineally back to Ukrainian Jewish[59] and Lithuanian Jewish women.

I found a non-Jewish Italian family carrying H1aj1 whose matriline traces back to a woman born in the mid-nineteenth century in the municipality of Cogoleto in the region of Liguria in northwestern Italy.[60] As of January 2022, Family Tree DNA's mtDNA Haplotree lists a total of three carriers of H1aj1 with matrilines from Italy.

Its parent haplogroup, H1aj, has been found in Syria, Denmark, Poland, Russia, and among the French.

H1aj1a

Ashkenazim with matrilines from Moldova, Ukraine, Hungary, Austria, Poland, Belarus, and Lithuania are among the carriers of H1aj1a. Compared to H1aj1, H1aj1a is more than twice as common among Ashkenazim.

A particular Ashkenazic carrier of H1aj1a who tested at Family Tree DNA matches a Portuguese person with a matriline from the city of Porto in the Full Coding Region screen with a genetic distance of 3.[61] The Portuguese person has also been assigned to the specific branch H1aj1a. For now, that is the closest known non-Jewish match.

Family Tree DNA's mtDNA Haplotree's H1aj1a listing includes a second Portuguese carrier plus a carrier with a matriline from Iraq and a carrier with a matriline from Ireland. It also has some carriers from Latvia, Romania, Slovakia, Germany, and the Netherlands but some or all of those are likely Ashkenazic.

YFull's MTree estimates that the common ancestor for all H1aj1a carriers in its database lived about 1,200 years ago but with a confidence interval that enables a date as early as 2,700 years ago. MTree also estimates that H1aj1a became distinguished from H1aj1 about 6,700 years ago through the accrual of four mutations: G207A, G9621A, T16172C, and G16456A.

H1as2

Among Ashkenazim, the mtDNA haplogroup H1as2 is most prevalent in Polish Jewish matrilines but also exists among Ashkenazim with matrilines from Germany, Austria, Czechia, Moldova, Romania, Ukraine, Belarus, and Lithuania.[62] H1as2 was also found in an eighteenth-century sample from the town of Hamina in southern Finland.[63]

Its parent haplogroup, H1as, has been found among the peoples of modern Poland and ancient Scotland. Reinforcing the European connections, there is a Portuguese H1as1a sample, a H1as1a1 sample from

Scotland, and H1as1a samples from the eleventh-twelfth centuries from Radom, Poland.

H1ax

H1ax is an extremely rare mtDNA haplogroup among Ashkenazim. It was not detected to be carried by any of the Ashkenazic participants in Costa's team's mtDNA research[64] but I found and received confirmation that some Ashkenazim in the nineteenth century did carry it and that they were not converts. Their descendants had their mitochondria completely sequenced.

Customers of Family Tree DNA who belong to H1ax have matrilines from countries including Ireland, England, France, Germany, Poland, and Hungary, and some of those lines are not Jewish. A person carrying H1ax whose matriline comes from Austria was studied by Anita Kloss-Brandstätter.[65] A carrier of H1ax whose matriline comes from Italy is also known.[66]

A female H1ax carrier who lived circa 400–200 B.C.E. was found in Kent, England.[67]

A daughter branch of H1ax that is called H1ax1 is found among non-Jews in England, Wales, Scotland, and Sweden. This reinforces the already known European associations of H1ax and makes it clear that the first Ashkenazic H1ax would have been a convert to Judaism who lived prior to the start of Ashkenazic genealogical paper trails.[68]

H1b

A relatively small proportion of Ashkenazim carry the root level of H1b, also called H1b*, as confirmed by full sequence testing of their mtDNA at Family Tree DNA. This is a widespread European haplogroup that has also been found in people in places including Finland, Bulgaria, and Sardinia and among the Basques. Ashkenazim with this precise haplogroup who tested with Family Tree DNA match some Europeans in their Full Coding Region screen, including non-Jewish Polish and Swedish people.[69] An archaeological sample from Poland that apparently dates to the Neolithic (New Stone Age) era and was uploaded to GenBank by Maciej Chyleński's genetics team carries H1b.[70]

H1b1a

Some Ashkenazim who tested their complete mtDNA sequences with Family Tree DNA and YFull belong to this branch of H1b. Some of them trace their matrilines to Lithuania. H1b1 is also found among the Druze people but it is most commonly found among Europeans both in modern and ancient times.

The Ashkenazic variety of H1b1 is defined in part by the T16362C mutation and therefore called H1b1-T16362C by Family Tree DNA but more precisely identified as H1b1-T16362C's daughter subclade H1b1a in YFull's MTree and Ian Logan's mtDNA database as of 2021 because the Ashkenazim also carry the extra mutation A8348G, as Wim Penninx had found in 2019.[71] Family Tree DNA's mtDNA Haplotree does not assign any of its customers to H1b1a as of 2021.

H1b1-T16362C is found in modern people in countries including Spain, Italy (including its central region of Umbria), Denmark, Sweden, Serbia, and Czechia, and also has been found among Basques. Additionally, H1b1-T16362C was found in pre-modern samples from Denmark, England, and the Baltic region.[72] That sample from England was found in London's St. Mary Spital cemetery among skeletons dating to between 1400–1539. The sample from Denmark comes from a rural parish cemetery "HOM 1046" in the village of Sejet near Horsens in the eastern Jutland region and it dates to between 1150–1574. None of those samples carry the mutation A8348G.

H1b1a, the exact branch that Ashkenazim possess, is also present in non-Jews in modern Italy,[73] Sweden, Hungary's Szeged region,[74] Poland,[75] and Russia (Novgorod region, Tatars in Tatarstan, and Buryats).[76] Some of these point to a likely European ancestor for Ashkenazim.

Other European non-Jews carry subclades of H1b1a defined by extra mutations that Ashkenazim do not carry. These include the mutations A10978G in Germany, A374G in Italy and Russia, C16185A in Denmark, and T13608C in Poland and the Netherlands. As a result, YFull's MTree places those samples into newly identified subclades it calls H1b1a1, H1b1a2, H1b1a3, and H1b1a4 respectively.

H1b2a

Some Ashkenazim carry H1b2a and its daughter branch H1b2a1. For example, H1b2a is present among Ashkenazim with matrilines from Ukraine, Belarus, Lithuania, Romania, and Hungary and H1b2a1 is present among Ashkenazim with matrilines from lands including those now in the Netherlands, Germany, Poland, Slovakia, and France (including the Alsace region bordering Germany). H1b2a was also found in a Mongol from Mongolia and in a YFull customer with a matriline of unlisted ethnicity from Russia.

H1b2a probably entered the Ashkenazic population from a European convert to Judaism. Its parent haplogroup, H1b2, is present among non-Jewish Russians as well as in countries like Poland, Czechia, Bosnia and Herzegovina, Serbia, Albania, Denmark, Sweden, and Finland. H1b2 has an old presence among Europeans. An ancient female carrier of H1b2 dating back to around 730–390 B.C.E. was located at Ķivutkalns, Latvia.[77] A H1b2 sample dating to the thirteenth-fifteenth centuries was found at Khiytola (Hiitola) in northwestern Russia's Republic of Karelia.[78] Another medieval H1b2 carrier was found at the "ØHM 1247" cemetery in Haagerup, Denmark.[79]

H1e4

Some full Ashkenazim who tested with Family Tree DNA have been assigned to the mtDNA haplogroup H1e4 (called H1e4* by YFull's MTree) and I confirmed that this is their complete sequence. At least one of them has an Ashkenazic matriline from Lithuania.

H1e4 probably emerged among non-Jewish Europeans. Its sister haplogroup H1e3 is found in France and Finland and among Welsh people, while H1e5a and H1e5b are found in Italy. H1e and downstream branches of it are found in Ireland, Scotland, Switzerland, Spain, and Portugal.

A number of ancient H1e Europeans have been unearthed. A male H1e was buried around 5210–5002 B.C.E. at the Halberstadt-Sonntagsfeld archaeological site in Germany, which was associated with the Linear Pottery culture.[80] A female H1e from the same period (5300–4900 B.C.E.) was excavated from the Pusztataskony-Ledence I site in

Hungary belonging to the Szakálhát division of the Alföld Lineal Pottery culture.[81] A male carrier of H1e1a dating back to around 3970–3710 B.C.E. was found in Esperstedt, Germany.[82]

H1e4a

Ashkenazim with matrilines from Germany, Austria, Hungary, Poland, Lithuania, Latvia, Belarus, Ukraine, Moldova, and Romania are among H1e4a's carriers.[83] It is very likely of European origin. Some of the Ashkenazim who tested with Family Tree DNA and belong to this haplogroup see an H1e4a carrier with a seventeenth-century matrilineal ancestor from Denmark in their Full Coding Region screen with a genetic distance of 1 away from themselves. Those same Ashkenazim are exact matches to a Dutch person with a nineteenth-century matrilineal ancestor with the distinctly Christian first name of Christina.[84] In addition, certain Family Tree DNA and YFull customers who are assigned to H1e4a have matrilines from Austria and Germany (including the Rhineland) that are not listed as Ashkenazic and might not be.

H1e4a carriers have a mutation called C16114T that H1e4 carriers lack. Ashkenazic H1e4a carriers and Ashkenazic H1e4 carriers do not show as matches to each other inside Family Tree DNA's mtDNA "Haplogroup Origins" screen even down to the HVR1 (Hyper Variable Region 1) level. It seems that separate women were responsible for the existence of these haplogroups in Ashkenazim.

H1f

The Eastern Ashkenazic mtDNA haplogroup H1f, which was inherited by some Jews in the Russian Empire, mostly Lithuanian Jews but also Ukrainian Jews, definitely has a northern European source. Ethnic Finnish people dominate the list of H1f carriers in Family Tree DNA at the root level H1f* and there are only a handful of carriers with non-Jewish matrilines from Sweden and Germany. The Ashkenazic carriers do not match these non-Jewish carriers in their Full Coding Region screen.

The T16093C mutation created a subgroup of this haplogroup called H1f-T16093C that exists among Family Tree DNA testers with matrilines from Sweden, Germany, England, and Scotland. H1f1, a daughter branch of H1f-T16093C, is very common among Finnish people in Finland and has a much smaller presence elsewhere in northern Europe but does also exist in Sweden and Norway and among Poles. There is a branch of H1f1 called H1f1a that is again also found mostly in Finland.

Ancient and medieval carriers of H1f and its branches were all found in northern Europe. An ancient H1f carrier found buried on southeastern Sweden's island of Öland comes from the Pitted Ware site of Köpingsvik and dates to between 5200–4850 B.C.E.[85] H1f1 was found in two medieval samples from Finland (Hollola and Tuukkala).[86] A male carrier of H1f1a who was part of the Wielbark culture was buried sometime between the years 80 and 260 C.E. at what is now Kowalewko in Greater Poland.[87] A much earlier H1f1a carrier, a female found at a Corded Ware culture's site in Karlova, Tartu, Estonia, dates back to circa 2440–2140 B.C.E.[88]

H1u2

H1u2, which requires the mutations G3483A and C16320T and was formed about 9,600 years ago according to YFull's MTree, is rare among Ashkenazim. It was not one of the H lineages found by Costa's team in 2013, being first identified as a potential Ashkenazic lineage in a preliminary survey by Jeffrey Wexler in 2018.[89] Inside Family Tree DNA, I found a full Eastern Ashkenazi who is assigned to H1u2 based on complete sequencing and also found several additional H1u2 carriers with some Ashkenazic ancestry, one with a Lithuanian Jewish matriline and another with a Ukrainian Jewish matriline. It appears to have also been present among Belarusian Jews.

I do not know whether or not the H1u2-carrying customers of Family Tree DNA who specified that their matrilines come from Czechia, Germany, Russia, and Finland are Ashkenazic. Definite non-Ashkenazic H1u2 carriers at Family Tree DNA include people with matrilines from countries like Italy, Greece, Azerbaijan, and Bahrain. The Greek carrier has a matriline from the island of Kos in the southeastern Aegean Sea.[90] In addition, H1u2 and its daughter branch H1u2a are found in Armenians from Turkey,[91] and the former is found in Danes. A Sardinian who had initially been placed at the ancestral level of H1 was later confirmed to be a member of H1u2 as well.[92]

Silva's team found two examples of H1u2 from Catalonia, Spain.[93] One of them shares the extra mutation A9563G with a Danish sample and YFull's MTree uses it to define what they call H1u2b.

Geneticists found two H1u2 carriers from commoner graves in Hungary from the tenth and/or eleventh centuries: one at the Homokmégy-Székes cemetery and the other at the Ibrány Esbóhalom cemetery.[94] YFull's MTree contends that these Hungarians belonged more precisely to the subclade H1u2a, which is defined by the presence of the mutation G8573A. Their MTree also places a modern sample from Boris Malyarchuk's population survey of Russia, presumably a non-Jew, who was formerly sequenced as H1u2, into H1u2a.[95]

I gathered mtDNA samples from two Crimean Karaites[96] and, although their mitochondria were only partially sequenced as H by Family Tree DNA, I discovered in 2022 that they belong to the Ashkenazic H1u2 cluster because they match those Ashkenazim in the HVR2 (Hyper Variable Region 2) screen. They also match the Azerbaijani in the HVR2 screen. The carriers with matrilines from Greece, Italy, and Bahrain do not match the Karaites in either the HVR2 or HVR1 screens. I saw that C16320T was identified in the Karaites at the HVR1 level[97] but that identifying the mutation G3483A evidently depends on complete sequencing.

H1bd

H1bd, a daughter subclade of H1, is a very rare haplogroup in Ashkenazim. It was not detected to be carried by any of the Ashkenazic participants in Costa's team's mtDNA research[98] but I confirmed that one full Ashkenazi got assigned to H1bd by Family Tree DNA as the result of complete mtDNA sequencing.

Other Family Tree DNA customers who belong to H1bd report their matrilineal ancestries to come from countries like Ireland, France, England, Switzerland, Spain, Mexico, Colombia, and Sweden according to the company's mtDNA Haplotree. In scientific research, H1bd has been found in people from Sardinia[99] and in Basques of northern Spain.[100]

H1bd was also present in pre-modern Scandinavia, as revealed by a sample from a grave at the rural cemetery "FHM 3970" in Nordby, Jutland, Denmark.[101]

YFull's MTree estimates that the common ancestor for all H1bd carriers lived in about 4850 B.C.E.

H1bo

Haplogroup H1bo is another Ashkenazic branch of H1. Some Ashkenazic matrilines with H1bo come from Germany, Poland, Ukraine, Belarus, and Lithuania. H1bo is also found in people with Greek Jewish matrilines that are either Romaniote or Sephardic and people from the Apulia region of southern Italy. Some of Family Tree DNA's Ashkenazic H1bo carriers match one of these non-Ashkenazic Jews with roots in Greece at a genetic distance of 1 in the Full Coding Region screen.[102] These same Ashkenazim also match people with matrilines from Italy and Turkey that are likely Jewish at a genetic distance of 1 and match definitely Jewish people with matrilines from Greece and Turkey at a genetic distance of 2.[103]

H1bo has also been detected outside of Jewish communities. YFull's MTree identified a H1bo carrier from, or with a matriline from, the city of Almería in Andalusia in southern Spain. Additional H1bo carriers with matrilines from Spain participate in Family Tree DNA's mtDNA Haplotree. As of the time of this writing, all three of Family Tree DNA's H1bo customers declaring matrilines from Spain match Ashkenazim in the Full Coding Region screen. One of these people with roots in Spain matches some of the Ashkenazim at a genetic distance of 1 while the other two (only one of whom explicitly lists a Sephardic matriline) match some of them at a genetic distance of 2. Agustina de Iniesta, who was born in 1732, was the matrilineal ancestor of a modern H1bo Spaniard from the Province of Almería[104] who is matching Ashkenazim with a genetic distance of 2.

Ian Logan confirmed to me that the company YSEQ assigned the haplogroup H1bo to a non-Jewish Italian person with a matriline from Taranto, the name of a city and province in Apulia.[105] The city of Taranto was home to Jews and "New Christians" of Jewish descent until both groups were expelled in the sixteenth century.[106]

I cannot tell for certain that the H1bo carrier who was tested by García-Fernández's team and listed as a non-Romany person from the "General population" of Romania[107] did not have a recent Ashkenazic ancestor. I also do not know whether or not the H1bo carriers from Austria, France, and the Netherlands who appear in Family Tree DNA's mtDNA Haplotree have Jewish matrilineal ancestors.

H1bw

H1bw is a branch of H1 that is so rare among Ashkenazim that it was not detected to be carried by any of the Ashkenazic participants in Costa's team's mtDNA research[108] but is carried by several bona fide full Ashkenazim at Family Tree DNA, including in a matriline from Belarus. Most of the Ashkenazic H1bw carriers in Family Tree DNA are exact matches (genetic distance = 0) in their Full Coding Region screen to people with matrilines from Lithuania, Poland, Ukraine, and Russia which are presumably the other Ashkenazic matches' ancestral locations but also to a non-Jew with a nineteenth-century matrilineal ancestor from France who had French first and last names of non-Jewish character.[109]

Outside of the Jewish and French communities, H1bw has been found in two Sardinians[110] by scientists. In Family Tree DNA, most Ashkenazic H1bw carriers have a genetic distance of 1 from an H1bw carrier with an eighteenth-century matrilineal ancestor with a Spanish surname, a genetic distance of 2 from H1bw carriers whose matrilines come from Spain, Cuba, Cape Verde, and Norway, and a genetic distance of 3 from H1bw carriers whose matrilines come from Portugal, Brazil, the United Kingdom, and Norway. Multiple Swedes carry H1bw, but they are even more genetically distant from the Ashkenazim, appearing as matches to them only at the HVR1 level.

I presume that H1bw was introduced to Ashkenazim by a converted European, possibly French, woman.

H2a1e1a

H2a1e1a is a very rare haplogroup among Ashkenazim. It was not found during earlier comprehensive investigations of Ashkenazic H haplogroups by Costa's team, Wim Penninx, and Jeffrey Wexler. In June 2021, I discovered that a fully Ashkenazic person with a Lithuanian Jewish matriline who had just recently tested with Family Tree DNA up to the Full Coding Region level got sequenced by that company into H2a1e1a and that a half-Ashkenazi with a Bessarabian Jewish matriline belongs to the same haplogroup.

The most frequent country of origin for H2a1e1a matrilines according to Family Tree DNA's mtDNA Haplotree is Germany, and at least one of those comes from Upper Franconia in Bavaria.[111] The Haplotree also shows carriers of this haplogroup with matrilines from Switzerland,[112] Denmark, Portugal, England, Ireland, France, the Netherlands, and other European countries (not including Poland, Italy, or Greece) plus one from Kazakhstan. Behar's team had likewise found this haplogroup in Ireland[113] and Li's team and Raule's team found five people from Denmark with it.[114]

An ancient carrier of H2a1e1a was found in Taforalt cave in northeastern Morocco's Berkane Province.[115]

A branch of it called H2a1e1a1 is found in modern Ireland and the United Kingdom.

Its parent haplogroup, H2a1e1, has a presence in Ireland and Denmark. H2a1e1 was also in a pre-modern (apparently Neolithic) sample from Poland that Maciej Chyleński's team uploaded to GenBank.[116] H2a1e1's parent haplogroup, H2a1e, is found in France, Denmark, and Scotland and in Bedouins from Israel as well as modern Germany and already was in Copper Age Germany.

My current impression is that H2a1e1a most likely came into the Ashkenazic population from a German convert to Judaism.

H2a2b1

The European mtDNA haplogroup H2a2b1 is rare among Ashkenazim but has been detected in bona fide full Ashkenazim at Family Tree DNA and 23andMe, including at least one with a matriline from Ukraine. H2a2b1 is also part of non-Jewish populations in countries like Estonia, Sweden, Denmark, France, and England. Ashkenazim who tested with Family Tree DNA have ethnic Swedish, German, and Scottish matches in their Full Coding Region screen[117] and for this reason I assume that the first Ashkenazic H2a2b1 was a northwestern European convert, specifically a German.

An eighth-century Viking male carrier of H2a2b1 was found at Salme on the island of Saaremaa in western Estonia.[118]

Its daughter haplogroup H2a2b1a1 exists in Scotland while its granddaughter haplogroup H2a2b1a1a exists in Denmark.

H3

Full mitochondrial sequence testing at Family Tree DNA established that H3 (also called H3*) is a haplogroup carried by some fully Ashkenazic people, including some with matrilines from Lithuania. Relative to H3*, haplogroups H3p and H3w are more frequent among Ashkenazim. At the Full Coding Region level, Ashkenazim with H3 match some European non-Jews, including one Italian, and they also match a Venezuelan whose matriline presumably came from Spain. YFull's MTree lists ethnic Ukrainian and French carriers of H3* as well as carriers from countries like Iceland and Finland. Family Tree DNA's mtDNA Haplotree lists numerous H3 carriers with matrilines from England, Ireland, France, Sweden, Norway, Germany, and Italy along with lesser numbers with matrilines from other European countries and regions including, but not limited to, Portugal, Belgium, Greece, Malta, Latvia, Serbia, and the Faroe Islands. Genetic studies found H3* carriers in Italy's Umbria and Sardinia regions and in Basques from Spain. But H3* is also found outside of Europe, including in Syria, Lebanon, Egypt, Algeria, and Morocco.

H3* was present in medieval Denmark[119] and England.[120] A H3 carrier was buried in the Phoenician necropolis of Puig des Molins on the Spanish island of Ibiza in the fourth century B.C.E.[121] A Copper Age female H3 carrier dating to around 3007–2871 B.C.E. was found underneath the city of Lucena in the province of Córdoba in southern Spain[122] and her haplogroup assignment of H3* rather than just H was confirmed by Ian Logan and by YFull's MTree. Three H3 carriers were discovered during archaeological digs in Switzerland.[123]

H3p

Substantial numbers of Ashkenazim with matrilines from Germany (including the Rhineland), Austria, France (particularly the Alsace region), Czechia, Hungary, Romania, Slovakia, Poland, Ukraine, Belarus, Lithuania, Latvia, and Russia are carriers of mtDNA haplogroup H3p. The solitary Catholic Pole with H3p who matches many of the Ashkenazim in the Full Coding Region screen at Family Tree DNA descends matrilineally from an Ashkenazic woman who, a baptismal record shows, became a Catholic in 1812.[124]

H3p is also present among non-Jewish Serbians.[125] A young man belonging to H3p was buried in the Viking Age mass grave in the churchyard of St. Lars Church in Sigtuna, Sweden.[126] These modern and medieval examples strongly suggest that Ashkenazim acquired H3p from a European convert to Judaism.

H3w

H3w is encountered among many Ashkenazim,[127] including some tracing their matriline to Lithuania and Poland and relatively fewer with matrilines from Belarus, Latvia, Germany, Austria, the Netherlands, Slovakia, Ukraine, and Romania. I do not know if the H3w carrier from France who gave a DNA sample to Behar's team is Jewish[128] but Leo Cooper's research confirmed that H3w is found in some Ashkenazic matrilines from France. Ashkenazic H3w carriers who tested with Family Tree DNA match some non-Jews in their Full Coding Region screen including a Spaniard with a matriline from the village of Rueda in the province of Valladolid in northwestern Spain and people with matrilines from Mexico and Argentina.[129] They also match non-Jews with matrilines from Thessaloniki, Greece and Ferrandina in southern Italy's Basilicata region.

Costa's team sampled three H3w carriers from Tunisia.[130] It is not stated whether those Tunisians are Jews, Arabs, or Berbers. The creators of YFull's MTree believe that the Tunisian samples belong to a different subclade of H3w than the European samples do. They tentatively call the Tunisian branch H3w2 and the European branch H3w1.

H3ap

Although H3ap was not one of the mtDNA H lineages identified as being present among Ashkenazim by Costa's team, I confirmed inside Family Tree DNA and GEDmatch that some full Ashkenazim do (but very rarely) belong to H3ap. For example, Ashkenazic matrilines from Romania and Lithuania have been sequenced as H3ap.

Evidence suggests that the first Jewish H3ap was a European woman who converted to Judaism. Most Ashkenazic H3ap carriers who have tested their mtDNA with Family Tree DNA see non-Jews within their Full Coding Region screen with a genetic distance of 3 away from themselves. Those matches include three people whose matrilines are Portuguese, including at least one from the Azores islands, and several who seem to have northwestern European matrilines.[131] This haplogroup is definitely found today among people in England, Scotland, Denmark as well as in Poland and Italy's Umbria region. It also exists among ethnic Bulgarian and Romany people.

Ancient H3ap carriers were found in archaeological sites in Switzerland[132] and England.[133] A daughter haplogroup of H3ap called H3ap1 was identified in a pre-modern sample from Denmark.[134]

H4a1a1

I confirmed that a fully Ashkenazic person was assigned H4a1a1 (also called H4a1a1*) by Family Tree DNA and that this is their completely sequenced haplogroup. It came to the Ashkenazic population from a different ancestor than the one who introduced H4a1a1a, discussed below.

H4a1a1 is also found in modern European non-Jews including from Finland, Norway, Sweden, Denmark, England, Wales, Ireland, Northern Ireland, and Scotland (in ethnic Scots and Orcadians). Family Tree DNA's mtDNA Haplotree additionally lists H4a1a1 carriers with matrilines from France, Italy, Germany, and the Netherlands.

A male carrier of H4a1a1 was found at the court tomb in Parknabinnia in County Clare, Ireland and his remains date to between 3642–3375 B.C.E.[135]

It is evident that Ashkenazim acquired H4a1a1 from a European proselyte.

H4a1a1a

This mtDNA haplogroup is rare in Ashkenazim but was found among Ukrainian Jews of Ashkenazic heritage.[136] Elsewhere, H4a1a1a is found in modern English, Scottish, and Uyghur people along with people in

Poland, Latvia, Finland, Denmark, Ireland, Bulgaria, and Italy's Umbria region. Inside Family Tree DNA's Full Coding Region screen, Ashkenazim with this haplogroup are exact matches to non-Jews from Germany, the United Kingdom, and Finland and are distant matches to non-Jews from Poland, Hungary, Norway, Denmark, and Russia.[137]

It had an equally widespread presence among European non-Jews in earlier centuries. A H4a1a1a-carrying female was buried in late-medieval times at the Cistercian abbey of St. Mary Graces in East Smithfield, London, England.[138] A H4a1a1a carrier was found at the rural cemetery of Nordby, Jutland, Denmark where the remains date to between 1050–1250.[139] A Viking male H4a1a1a who died between the tenth-twelfth centuries was found at Varnhem, Skara, Sweden.[140] Three H4a1a1a carriers were found buried in village cemeteries from Hungary from the tenth-eleventh centuries.[141]

Ancient carriers of this haplogroup have also been identified. An Early Bronze Age female H4a1a1a who died circa 1945–1777 B.C.E. was buried at the Wehringen-Hochfeld site in the Lech River valley in Germany.[142] Two H4a1a1a carriers from central Europe's Bronze Age Únětice culture are known. One is a male found under Prague, Czechia who died circa 1882–1745 B.C.E.[143] Another is a female found near Eulau, Germany who died circa 2132–1942 B.C.E.[144] A female H4a1a1a from the Early Neolithic era who died between 3940–3703 B.C.E. was found underneath the Poulnabrone dolmen (portal tomb) in County Clare, Ireland.[145]

As with its ancestral haplogroup H4a1a1, daughter branches of this haplogroup, including H4a1a1a1a, H4a1a1a1a1, H4a1a1a2, H4a1a1a3, and H4a1a1a4, are also generally distributed throughout Europe.

The first Jewish H4a1a1a was certainly a European convert, probably a German.

H4a1a3a

The mtDNA haplogroup H4a1a3a has a confirmed presence in a very low proportion of Ashkenazim, including those tracing their direct maternal lines to Ukraine. Ashkenazim with this haplogroup who did mtDNA testing through Family Tree DNA match several non-Ashkenazim of interest in their Full Coding Region screen, including a person with a Moroccan Jewish matriline from Casablanca as well as a person with a Spanish matriline from Riosa, Asturias.[146] H4a1a3a is also carried by a member of Family Tree DNA's "H4 mtGenome" project who had a Moroccan Jewish matrilineal ancestor from Meknes named Ester Dadoun.[147] That project also has H4a1a3a-carrying members with ethnic German and Norwegian matrilines.

H4a1a has been reported to be the second most common maternal haplogroup among Moroccan Jews tested by 23andMe but that company does not fully sequence all mtDNA haplogroups. As stated, H4a1a3a would be the terminal subclade for most or all of them.

H4a1a3a has also been confirmed by genetic genealogists to have been the lineage of a non-Jewish French woman named Marie Buisset who was born in Alsace-Lorraine in northeastern France in the 1670s.[148] Older than that is the female H4a1a3a-carrier who was dug up from the Christian cemetery Bensinstationen in Sigtuna, Sweden where at least some of the people were buried around the late ninth to early eleventh centuries.[149]

Behar's team found the subclade H4a1a3a1 in a modern person tracing matrilineal ancestry to Poland.[150] This person is apparently not Ashkenazic.

The parental haplogroup H4a1a3 has been found in modern matrilines from Ireland and the United Kingdom. Reaching further back on the haplotree, the ancestral haplogroup H4a1a exists in Italy, including the Umbria and Emilia-Romagna regions, as well as in Spain and Denmark and among German and British people. H4a1a was in ancient Switzerland.

H5

A very small proportion of Ashkenazim, including one with a matriline from Moldova,[151] carry the undifferentiated mtDNA haplogroup H5, which is sometimes differentiated from its descendant branches by referring to it as H5*. In their Full Coding Region screen in Family Tree DNA, I saw that they match matrilineal descendants of Christians from northwestern Europe, including Irish and apparently British women. Some of them have genetic distances of 2 and 3 from those Christians. This pattern appears to demonstrate that H5* entered the Ashkenazic population when a northern European woman married a Jew, even though H5* is also found among some non-Jews in the Middle East and the Caucasus.

H5-C16192T

This variety of the mtDNA haplogroup H5 is defined by the presence of the mutation C16192T, hence the suffix in the haplogroup's name as designated by Family Tree DNA. H5-C16192T is relatively uncommon among Ashkenazim but has been verified as the complete mitochondrial sequence in certain Jewish matrilines from Belarus,[152] Poland, Lithuania, and Ukraine. I do not have any other solid information about it at this time. A Neolithic-era sample from Turkey potentially belonged to H5-C16192T but scientists are unsure about that assignment.

H5-T16311C!

This is a distinct variety of H5 that is defined by the back mutation T16311C! and that is why Family Tree DNA attached that suffix to the haplogroup's name. H5-T16311C! is found in a tiny proportion of Ashkenazim, including those with matrilines from Romania and Poland, who had their complete sequences tested.[153] YFull's MTree has given H5-T16311C! the new nomenclature H5-c and designates the Ashkenazic branch of it, defined by the mutation C16167T, as H5-c1.

Five carriers of H5-T16311C! in the Family Tree DNA database have matrilines from Germany while smaller numbers have matrilines each from Austria, Russia, Italy, England, and Sweden. Some of these are not Jewish, including the Swedish customer Dick Wåhlin, who posted to his public blog the details of his direct maternal descent from the Swedish woman Magdalena Månsdotter who was born in 1780 to Lena Svensdotter who was born in 1737.[154] These Italians come from the general population of central and northern Italy and were assigned by YFull's MTree to non-Ashkenazic subclades that they call H5r*, H5r1, H5r2, H5s*, and H5t.[155]

A carrier of H5-T16311C! who is specified as belonging to the newly identified haplogroup H5-c2 by YFull's MTree participated in Behar's team's study that tested the mtDNA lineages of "Basque and non-Basque individuals from the Basque Country and immediate adjacent regions."[156]

A thorough examination of the pattern of matches within Family Tree DNA demonstrated that the Ashkenazic cluster is tight and nested within the larger European cluster. Some of the Ashkenazim match each other on their Full Coding Region screen exactly and others with a genetic distance of 1, whereas most or all of the non-Jewish matches are more distant with some of them displaying genetic distances of 2 and 3 from Ashkenazim and others being too distant to be reported as matches on the Full Coding Region screen at all.

H5a1

The mtDNA haplogroup H5a1, a descendant of H5's branch H5a, is rarely found in Ashkenazim. I received information that it is found in at least one Ashkenazic Jewish matriline from Ukraine and that non-Jewish European mtDNA matches are very noticeable to the Ashkenazic testers. I confirmed that it is their complete mitochondrial sequence. This haplogroup is evidently not shared by German Jews nor Sephardic Jews and this, combined with its scarcity in Ashkenazim overall, suggests a Polish convert to Judaism as the source.

H5a1 is otherwise a very common and widespread haplogroup. In Family Tree DNA and published studies, H5a1 is found in non-Jewish populations on three continents. In Europe, non-Jewish Dutch, French, Italian, Danish, Swedish, German, Polish, and Russian people are among its carriers along with people with matrilines from Serbia and Spain. In Asia, Buryat, Uyghur, Pamir highlanders, Iranians, and Kurds and people in India are H5a1 carriers. The known North African carrier is a Mozabite Berber from Algeria.

Unsurprisingly, a significant number of H5a1 samples have also been recovered from ancient and medieval Europe and Asia as far west as Scotland and as far east as Siberia and Xinjiang. For example, a male H5a1 carrier from medieval Hungary was unearthed from a commoner grave.[157] In 2014, Anna Juras' genetics team noted that three H5a1 carriers dating back to the Roman Iron Age had been found in Poland, specifically at the Kowalewko, Gąski, and Rogowo burial sites, as well as one H5a1 from Poland during the Middle Ages and they noted that the fact that this haplogroup is also found in modern Poles "suggests a genetic continuity of certain matrilineages in the territory of present-day Poland" spanning the pre-Slavic through Slavic eras.[158] In 2020, two more H5a1 samples from pre-modern Poland were uploaded to GenBank by Maciej Chyleński's team.[159] An adult female carrier of H5a1 was buried sometime between 2575–2349 B.C.E. in a grave at the Sope burial site of the Corded Ware culture under the village of Jabara in northeastern Estonia.[160] Two H5a1-carrying female members of the Bell Beaker culture were buried in Bavaria in southern Germany circa 2500–2000 B.C.E. and one other one who died between 2118–1937 B.C.E. was buried in North Holland.[161]

Descendant branches of H5a1 are common across Europe. For instance, H5a1a is found in ethnic Polish and Dutch people, H5a1c1a in Danes, and H5a1f in Scots, Czechs, and Poles, among many others.

H5a, its subclades H5a2, H5a3, H5a5, H5a4, H5a6, and H5a8, and their descendants, are present among many European Christian ethnicities, including inhabitants of Sweden, Norway, Ireland, England, Poland, Greece, Italy, Spain, and many other countries. The next entry indicates the kinds of European Christians who carry its other subclade, H5a7. H5a is absent among Middle Eastern non-Jews.[162]

H5a was found in the remains of a man from the Tagar culture who was buried in the Abakano-Pérévoz II archaeological site between circa 800 B.C.E. and 100 C.E.[163] This site is located in southern Siberian Russia's Bogratsky region within Khakassia, and this man shows signs of being of mixed European-East Eurasian origins, but most other members of this culture came from fully European genetic origins. Among the data available to the team that studied this man's genetics, his mtDNA haplogroup reflected the European portion of his heritage, since the specific haplotype of H5a that he carried was found among modern populations only in a single Austrian.[164]

H5a7

A small proportion of Eastern Ashkenazim carry the mtDNA haplogroup H5a7 which is found among Ashkenazim with matrilines from places including Belarus, Ukraine, and Poland.[165] A particular Ashkenazi with a matriline from Poland whose sample[166] was collected by Behar's team has been confirmed to belong to its root level H5a7* rather than one of the newly identified subclades of it that have been researched for YFull's MTree.

H5a7 is also found among non-Jewish matrilines from Poland, Switzerland, Denmark, Norway, Scotland, Ireland, Wales, Spain,

Portugal, and Italy (including the Molise region). Of these, some of the carriers with matrilines from Switzerland, Denmark,[167] and Spain[168] were confirmed to be fully sequenced as H5a7* by MTree whereas at least one Italian is in the newly named subclade H5a7a and at least one Norwegian is in the newly named subclade H5a7b. In the Full Coding Region screen, the closest non-Jewish matches to the Ashkenazic H5a7 carriers who tested with Family Tree DNA have matrilines from Poland and Bollingen, Switzerland and both of them have genetic distances of 1 from Ashkenazim. The next closest non-Jew, with a genetic distance of 2 from Ashkenazim, has a matriline from Maastricht in southeastern Netherlands. A non-Jew with a matriline from Poschiavo, Switzerland has a genetic distance of 3 from Ashkenazim.[169]

A medieval carrier of H5a7 was buried in a grave at the rural cemetery "FHM 3970" in Nordby, Jutland, Denmark.[170]

At this time, one cannot assume that H5a7 came into the Ashkenazic population from a northern, central, or eastern European woman rather than a southern European woman.

H5c2

The mtDNA haplogroup H5c2 has been found in many Ashkenazim with matrilines from Germany,[171] Austria, Hungary, Romania, Slovakia, Poland, Ukraine, Belarus, and Russia proper. Family Tree DNA's Full Coding Region screen indicates that at least some of these Ashkenazim have a genetic distance of 1 from a non-Jew with a matriline from colonial South Carolina and a genetic distance of 2 from a matriline from Klein-Umstadt in Germany's Hesse region.[172] The first Ashkenazic H5c2 could have therefore been a German convert to Judaism.

Its parent haplogroup, H5c, is found in Iran. Another branch of H5c, called H5c1a, is found in Finland.

H6a1a1a

H6a1a1a is the mtDNA haplogroup of some Ashkenazic matrilines from Poland, Lithuania,[173] and Romania. Some Ashkenazim who tested their mtDNA at Family Tree DNA up to the Full Coding Region level have a genetic distance of 2 from a Turkish Jew from Izmir and a Turkish Jew with a matriline from Gallipoli.[174] They also have a genetic distance of 1 from a Brazilian non-Jew with a matriline from Rio de Janeiro and a genetic distance of 2 from a Colombian non-Jew with a matriline from the town of Bugalagrande. Other H6a1a1a carriers at Family Tree DNA have matrilines from North Macedonia, Algeria, Tunisia, and Spain and those are potentially from Sephardic Jewish ancestors, but I do not know whether or not those people are Jewish or descend from Jews. A non-Jew from Colombia who belongs to H6a1a1a is a direct matrilineal descendant of Isabel Álvarez de Alcocer, who was born around 1500 in the village of Medellín in western Spain's Extremadura region.[175] Costa's team found an Iranian Jewish carrier of H6a1a1a.[176]

The evidence suggests that Ashkenazim inherited H6a1a1a from a Middle Eastern ancestor. Costa's team called its parent haplogroup H6a1a1 a "Near Eastern subclade."[177] Today, H6a1a1 is found in places including the United Arab Emirates, North Ossetia, Greece, Italy, Libya, and Hungary. A Syrian Jew and a person with a Sephardic Jewish matriline from Turkey were assigned H6a1a1 by 23andMe but that company does not fully sequence all haplogroups, so it is possible that they are really other H6a1a1a carriers.

The ancestral haplogroup H6a1a has a wide distribution with samples from Saudi Arabia, Spain, Spain's Canary Islands, Bulgaria, Italy's Umbria region, France, but also Denmark, Ireland, Sweden, and Czechia. Its geographical distribution was also widespread in ancient times. For example, a Bronze Age H6a1a female was buried in the De Tuithoorn site in the Netherlands sometime between 1883–1664 B.C.E.[178] and a H6a1a female from the Kangju culture was buried between the years 264 and 273 in a kurgan at the Kok-Mardan site in southern Kazakhstan.[179]

H6a1a3

An Ashkenazi with a matriline from Odessa, Ukraine whose GenBank entry[180] lists the haplogroup H6a1a has been confirmed by Ian Logan to belong more precisely to H6a1a3. That sample originated with Family Tree DNA and inside that company's database, I independently confirmed that multiple full Ashkenazim carry H6a1a3, as sequenced by that company, and also include one with a matriline from Hungary. According to Wim Penninx, these Ashkenazim have the mutations called A16482G and C16519T and they closely match at least one non-Ashkenazi inside Family Tree DNA.

As of January 2022, dozens of other Family Tree DNA customers with H6a1a3 have matrilines from European countries—England (23 participants), Scotland (8), Ireland (11), Germany (16), the Netherlands, and Finland (13)[181]—and there are no Middle Eastern or Asian carriers listed. Most of them had Christian first and last names.[182]

According to Family Tree DNA's mtDNA Haplotree, its daughter subclade H6a1a3a exists in matrilines from England, Scotland, Ireland, Northern Ireland, and France.

A female H6a1a3 carrier who fell victim to the Black Death of the fourteenth century was unearthed from the mass burial ground at East Smithfield, London, England.[183]

It is evident that the first Jewish H6a1a3a was a European convert.

H6a1a5

A sizeable number of Eastern Ashkenazim, including Jews with matrilines from Moldova, Ukraine, Slovakia, Poland, Lithuania, and Belarus, possess the European mtDNA haplogroup H6a1a5.[184] It is not found in German Jews. The closest definite non-Ashkenazic match to the Ashkenazim who tested with Family Tree DNA is a person from Sweden. This haplogroup has also been found in Finnish people and a non-Jew from Slovakia. A male from a Viking Age site in Ribe, Jutland, southern Denmark who lived around the ninth-eleventh centuries was a carrier of H6a1a5.[185]

A branch of this called H6a1a5a is found among non-Jews in Russia's Ryazan Oblast.

As with H6a1a3 and H6a1a5, some of their sister haplogroups have European distribution patterns. H6a1a4 is found in Finland, H6a1a8a is found in Wales, H6a1a9 is found in Denmark, and H6a1a10 is found in Sweden.

H6a1b2

A very small proportion of Ashkenazim carry the mtDNA haplogroup H6a1b2. Outside of the Ashkenazic community, most H6a1b2 carriers are non-Jewish Europeans. Family Tree DNA's mtDNA Haplotree lists many H6a1b2 participants with matrilines from countries including, but not limited to, European countries like Ireland, England, Sweden, Scotland, Germany, Finland (in ethnic Finns), Portugal, Italy, and France but also one from Morocco and one from Turkey. Some of the modern H6a1b2 carriers in GenBank have matrilines from Spain, Norway, Denmark, Russia, and India's Bihar state.

Pre-modern H6a1b2 samples have been located in Great Britain. A victim of the Black Death who was buried in the mass burial ground at East Smithfield, London, England sometime between 1348–1350 was found to belong to H6a1b2.[186] An H6a1b2-carrying male who was either a gladiator or a soldier was buried at Driffield Terrace in England's Yorkshire region during the Roman era.[187] A male H6a1b2 was buried in Cornwall circa 754–416 B.C.E.[188] A Bronze Age male H6a1b2 carrier who died between 1441–1272 B.C.E. was found on the island of Pabay Mòr in western Scotland.[189]

Branches of H6a1b2 also have a northern European distribution pattern. Haplogroups H6a1b2a and H6a1b2b are found in England, H6a1b2h in Scotland and Ireland, H6a1b2e in Ireland and Denmark, H6a1b2f in Denmark, and H6a1b2c in Germany.

H6a1b2's parent haplogroup, H6a1b, has been found in Bronze Age samples from Poland, Russia, and England.[190] H6a1b is found today in European countries (e.g., Ireland, Scotland, Hungary, Ukraine, and Bulgaria) and Middle Eastern ethnicities (e.g., Armenians from Turkey, Palestinian Arabs, and Druze) as well as further east.

H6a1b3

A small minority of Ashkenazim with matrilines from the Russian and Austro-Hungarian empires, including from Latvia, carry the mtDNA haplogroup H6a1b3. Among non-Jews, H6a1b3 has been found among many modern European ethnicities. Some of its haplotypes are specifically West Slavic: one is Czech and another two are Polish.[191] Other modern H6a1b3 carriers have non-Jewish matrilines from countries including, but not limited to, Ireland, Scotland, England, France, Italy, Switzerland, Denmark, and Sweden. The solitary Italian matriline came from the Campania region of southwestern Italy, distant from all of the other matrilines' home regions.

In Family Tree DNA's Full Coding Region screen, several of the Ashkenazim have genetic distances of 2 and 3 from all of their non-Jewish matches, and match two fellow Ashkenazim exactly and nine Ashkenazim with a genetic distance of 1 but there is one Ashkenazic outlier at 3, probably indicating unique mutations in that person's recent matriline. The Italian match is 3 steps distant from my Ashkenazic correspondent, whereas several of the 2-step matches' matrilines were German Christian, making it more likely that the source of H6a1b3 for the Ashkenazic population lived north of Italy.

H6a1b3's branches H6a1b3a and H6a1b3b are found among many kinds of non-Jewish Europeans; for example, both branches are found among Germans and the English.

A male who was buried circa 1930–1750 B.C.E. in the Ölljsö megalith in Sweden carried H6a1b3.[192] H6a1b3 was also carried by two members of the Únětice culture who died circa 2200–1550 B.C.E. in Leau in central Germany.[193]

H7c2

The mtDNA haplogroup H7c2 is found among Ashkenazim with matrilines from Ukraine, Belarus, and Hungary, for instance. In the Full Coding Region screen at Family Tree DNA, an Ashkenazic H7c2 carrier matches one Dutch person and one Welsh person.[194] Some of the H7c2 carriers who tested with YFull have matrilines from Wales, Romania, and central Italy's province of Massa-Carrara in the Tuscany region. Nicola Raule's scientific team found a H7c2 carrier with a matriline from Italy's Calabria region.[195] A Sardinian carrier of H7c2 was sampled by Anna Olivieri's team.[196]

The evidence we have suggests that H7c2 entered the Jewish population from a European proselyte. Leo Cooper believes that an Italian woman, rather than a northern European woman, was the more likely source of H7c2 in Ashkenazim.

H7c2's sister clade H7c1 (unlike the other sister clade, H7c3) includes some Middle Eastern members, including Druze and Saudi Arabian people, so it is not as exclusively European as the other two Ashkenazic branches of H7. However, H7c2's parent haplogroup, H7c, would still appear to have originated in Europe. The earliest known example of a H7c carrier, a male dating to between 5641 and 5560 B.C.E., comes from an archaeological site in Kargadur, Croatia that is representative of the Neolithic-era Cardial Ware culture.[197] Modern H7c carriers who participate in Family Tree DNA's mtDNA Haplotree include people with non-Jewish matrilines from Sweden, Norway, Germany, England, and Italy. An Algerian Jew who tested with 23andMe was assigned to H7c but could belong to a subclade of it because that company does not do complete mtDNA sequencing.

Most other ancient examples of H7 subclades similarly come from central and southeastern Europe, including an H7a1 sample from Bulgaria from circa 4711–4542 B.C.E., an H7d sample from Czechia from circa 2900–2200 B.C.E., an H7d5 sample from central-east Germany from circa 3700 B.C.E., and an H7h sample from Germany from circa 2000 B.C.E.

H7 has an ultimately European origin. A male member of the Eastern Linear Pottery culture who was buried at the Tiszadob-Ókenéz archaeological site in eastern Hungary circa 5300–4900 B.C.E., during the Middle Neolithic era, was determined to belong to undifferentiated H7.[198] A female who was buried at the Tell Yunatsite archaeological site in southern Bulgaria circa 4455–4359 B.C.E., during the Chalcolithic era (Copper Age), likewise carried undifferentiated H7.[199] A female carrier of H7 who was associated with the Middle Eneolithic period's Lasinja culture was among the dozens of victims of a massacre around 4180 B.C.E. in what is today Potočani, Croatia.[200] It comes as no surprise that H7 continued to have a presence in early medieval Hungary and is also found among modern Hungarians and among nearby ethnic groups including Slovaks and Romanians.

The first woman with H7 probably lived in the southeastern corner of Europe and her progeny married descendants of Anatolian farmers.

H7e

H7e is another mtDNA haplogroup descended from H7. H7e is found among Ashkenazim with matrilines from Germany, Austria, Romania, Moldova, Poland, Belarus, Lithuania, Ukraine, and Slovakia and it is more frequent among Ashkenazim than H7c2 and H7j. Its subclades H7e2 and H7e3 developed among some of these Ashkenazim and they are defined by the mutations G8994A and G12651A respectively.

Other H7e carriers have known matrilines from non-Jewish Europeans, including four from Germans, one from an inhabitant of Croatia's island of Susak, one from Albania, and three descending from colonial American families of European Christian heritage.[201] Most of these non-Jews' varieties of H7e have multiple unique mutations not shared by the Jews and they present indications that a long-term separation had occurred between the Jewish and non-Jewish branches, similarly to the vast majority of other Ashkenazic matrilines. The non-Jews' H7e varieties also contain more genetic diversity than the Jewish H7e branches do. Nevertheless, the Ashkenazic H7e carriers who tested their mtDNA with Family Tree DNA do see several ethnic Germans and several Poles matching them.[202] Some ethnic German H7e carriers have the mutation G8027A and are therefore placed into H7e1.

Based on H7e's modern distribution and the known aspects of Ashkenazic migrational and settlement history, Doron Yacobi and Felice Bedford discussed possible chronologies for when the H7e-carrying woman married an Ashkenazic or proto-Ashkenazic man. If that woman was a German, she probably lived during the ninth or tenth century, rather than during the seventh or eighth century.[203] They also surmised that H7e could have actually come from southern Europe and that the woman was either from Italy or southern France, even though H7e had not yet been found in those areas at the time of their study.[204] They believed that H7e could predate the period of Ashkenazic settlement and pointed out that the absence of H7e in modern Italy and France would not necessarily mean that ancient or medieval populations in those areas never carried it, since haplogroups sometimes die out in certain populations. Subsequent to their study, a H7e carrier from Sardinia appeared in the data set for a study by Olivieri's team.[205]

Although we do not know the tribe or ethnicity that the first Jewish H7e had been born into, it is evident that H7e had an ultimately European origin. For now, I will assume that the first Jewish H7e women came from a German family.

H7e was also in at least one population in western Siberia. A male H7e carrier was laid to rest under a mound at the Sargat culture's burial ground in Bitiya, Russia, north of Kazakhstan, sometime between 403–257 B.C.E.[206]

H7j

A branch of the mtDNA haplogroup H7 with the mutation T1700C has tentatively been named H7j. It is found among Ashkenazim with matrilines from Poland and Austria.[207] It is one of the three Ashkenazic H7 lineages discussed in a genetic study by Yacobi and Bedford.[208] Although Yacobi and Bedford characterized H7j as an "exclusively Ashkenazi Jewish" branch without "any non-Jewish affiliation," it has close non-Jewish matches who did not elect to upload their complete mtDNA sequences to the public GenBank database and partly for that reason were absent from their study.

H7j has a daughter lineage called H7j1 that is defined by the presence of the extra mutation T11137C. H7j1 is found in Ashkenazim with matrilines from Ukraine, the Crimea, Poland, Czechia, Belarus, and Lithuania.[209] As an Ashkenazic carrier of H7j1, I also have access to the list of matches and have received communications from some of them, and will present an anonymized summary of their ancestries for the first time in print here. On the Full Coding Region screen, my exact matches and my matches with a genetic distance of 1 away from myself are all Ashkenazim, while my matches with a genetic distance of 2 from myself include Ashkenazim plus two people with ethnic German maternal roots. One of those is a person for whom Family Tree DNA's MyOrigins admixture calculator reports an absence of Jewish autosomal DNA who has a Methodist Christian German matrilineal ancestress from the mid-nineteenth century. That ancestress had roots from northern and/or northeastern Germany, areas formerly part of Prussia, and she and her mother married other Germans, connected to the Holstein and Mecklenburg regions respectively. They spoke Low German. This person does not have the mutations T11137C (defining H7j1) and T1700C (defining H7j) and therefore is only in H7. My other German match at this genetic level descends from many non-Jewish families and has a nineteenth-century matrilineal ancestress whose surname is German Christian and was never carried by Ashkenazim.

My Full Coding Region matches with a genetic distance of 3 are all non-Jews. At least one of them has a matrilineal ancestress who was a German Christian. Another's matrilineal ancestress was a Dutch Christian woman who was born in 1807 in the city of Den Helder in northernmost North Holland province. A third match's matrilineal ancestress had a common Christian first name and lived in the late seventeenth century in the English colony of New York where she married a man of Dutch Christian origin with roots in the New Netherlands colony. A fourth match is an ethnic Swede whose matriline had all Swedish women back through the early 1700s. Lännäs, Örebro County, central Sweden was the birthplace of this Swede's matrilineal ancestress who was born in 1729. The Swede lacks the mutations T11137C and T1700C. The Swede has a genetic distance of 1 from some of the other Ashkenazim as well as 1 from the aforementioned Methodist North German descendant and a genetic distance of 2 from the Den Helder Dutch descendant.

My more distant matches continue the observed West Germanic pattern but also include one North Germanic outlier. My matches at this next level, in the HVR2 screen, include many additional non-Jews with matrilineal ancestresses from German Christian families, including several that definitely come from Prussia, plus one Swede whose matrilineal ancestress was a Swedish Christian in southern Sweden in the early nineteenth century.

Reaching back even further upstream from the H7j branch are people directly descended in their maternal lines from Dutch Christian and Norwegian Christian ancestresses. They share all but one (573.1C) of the HVR2 mutations of my closer matches, in addition to all of their HVR1 mutations. The Germanic pattern gives way to a general European spread beyond this point. An H7 individual matrilineally descended from a Slovak Christian woman shares all but two of my HVR1+HVR2 mutations. My even more distant HVR1 matches carrying haplogroup H7 include people with maternal French, Polish, English, Norwegian, Albanian, and Bulgarian roots, among other ancestries.

The most likely contributor of H7j to the Ashkenazic population was a North German woman of Saxon origin, rather than a Dutch, Frisian, or Scandinavian woman.

Note that the so-called "H7j" haplogroup in YFull's MTree is not the same as my aforementioned branch of H7. As of the time of this writing, the root level of H7j is not separately shown anywhere on MTree, while H7j1 is called H7o on it and YFull estimates that the most recent common ancestor of H7o lived around 500 C.E. That estimate might perhaps be skewed by the omission of its parent haplogroup.

The person from Denmark who is included in MTree's list of H7o samples probably had an Ashkenazic matriline because their sample comes from a study of Denmark's population that, based on what Felice Bedford told me,[210] might not be limited to ethnic Danes.[211] But if that person is an ethnic Dane or other type of Germanic gentile, H7j and H7j1 must have had independent contributors to Ashkenazim. Public databases do not reveal that person's ethnicity.

A particular person whose recent ancestry is all Turkish Jewish was assigned to haplogroup H7 by 23andMe but that company does not indicate what branch of H7 this might be or what mutations are included; they assign Ashkenazim who belong to H7e or H7j to simply H7. The Turkish Jew could not be in the branch H7c2 since Ashkenazic members of H7c2 are placed into H7c by 23andMe. The Turkish Jew could potentially belong to either H7e or H7j but could belong to another branch entirely (perhaps H7b or H7b1, which are found in some non-Ashkenazic Jewish lineages from the eastern Mediterranean region) if this was not the result of having a distant Ashkenazic ancestor who assimilated into a Sephardic community in Turkey.

H10a1b

Among customers of Family Tree DNA, the mtDNA haplogroup H10a1b has a presence among Ashkenazim with matrilines from Ukraine.[212] H10a1b is also found in customers of YFull with matrilines from Lithuania and Israel whose ethnicities are not stated and those two have the mutation T7220C that at least one person with a matriline from Ukraine lacks but which another such person has (along with the extra mutation G7356A). YFull's MTree estimates that the H10a1b carriers from Ukraine, Lithuania, and Israel descend from a common ancestor who lived around the year 750.

Outside of the Ashkenazic community, H10a1b is also found in Portugal. A person who lived around 300 B.C.E. at the Late Iron Age settlement called Çemialo Sırtı, excavated in Batman province in southeastern Turkey, was a carrier of H10a1b.[213]

Its parent haplogroup, H10a1, has a wide range, being found among present-day inhabitants of such places as Spain, Italy, Germany, Poland, Georgia, Iran, and Tibet.

H11a1

This mtDNA haplogroup is found at a low frequency in Ashkenazic populations. It has been found in Ashkenazic matrilines tracing back to Ukraine, Bessarabia, and Romania. It is otherwise spread across a number of European ethnicities including Poles, Lithuanians, Finns, Swedes, and people in Italy, Austria, Denmark, Slovakia, and Russia (including, but not limited to, among Volga Tatars). Some of these non-Jews are exact matches to Ashkenazim in the Full Coding Region screen while others have genetic distances of 1, 2, or 3.[214]

A medieval person whose haplogroup was H11a1 was buried in the Capidava necropolis in the province of Dobruja in southeastern Romania.[215] Another medieval H11a1 carrier, whose remains date to between 1200 and 1400, was found at Tuukkala, Finland.[216]

The daughter haplogroups H11a1a and H11a1a1 are especially found among Poles, although H11a1a has also been found in Austria and Serbia. In light of H11a1 itself also being present in Poles, I propose that the first Jewish H11a1 was a Polish convert.

H11a2a2

A small proportion of Eastern Ashkenazim with matrilines from Poland,[217] Ukraine, and Lithuania, carry the Balto-Slavic mtDNA haplogroup H11's subclade H11a2a2, which was formed circa 1100 according to YFull's MTree, with an earliest possible formation circa 850. Among Christians, H11a2a2 peaks in frequency in Poland, but it is also found in nearby countries like Ukraine and Belarus as well as in several other countries including Slovenia, Greece, Romania, Russia, and Finland.[218] Some Ashkenazic H11a2a2 carriers who have tested their mtDNA with Family Tree DNA see non-Jews within their Full Coding Region screen with genetic distances of 2 and 3 away from themselves, including Poles, Russians, a Finn, and a Bulgarian, and a Greek, with the number of Polish matches predominating over the other kinds.[219] A Pole was probably the ancestor of this haplogroup for Ashkenazim.

Sometime after the Black Death but before the seventeenth century, a H11a2a2 carrier was buried in East Smithfield, London, England.[220]

The connection of H11a lineages to the Balto-Slavic lands reaches deep into time. H11a was the haplogroup of a male member of the Narva culture whose remains, dating to approximately 4440–4240 B.C.E., were found at Spinigas, Lithuania.[221]

H11b1

The mtDNA haplogroup H11b1 has been found among Ashkenazim with matrilines from Romania,[222] Austria, Poland, and Ukraine. As is the case with H11a1 and H11a2a2, the first Jewish H11b1 carrier appears to have been a converted Slavic woman since H11b1 peaks in frequency among modern Ukrainians. Some of these Ashkenazim who tested their mtDNA with Family Tree DNA up to the Full Coding Region level match a non-Jewish person with a matriline from Czechia with a genetic distance of 2 away from themselves and non-Jews with non-Jewish matrilines from Czechia, Serbia, and Finland (specifically ethnic Finns) with a genetic distance of 3.[223] This haplogroup is also found among Danes,[224] Poles,[225] Roma people from Ukraine,[226] and Estonians. Its parent haplogroup, H11b, is found in Finland.

A Western Scythian H11b1 carrier who was found under Kup'evaha village in the Kharkiv region of eastern Ukraine died between 750–405 B.C.E.[227]

H13a1a1

H13a1a1 is rare in the Ashkenazic population. It was found in at least one Ashkenazic matriline each from Hungary and Ukraine. It has also been found in two full Iranian Jews who tested with Family Tree DNA.[228] Otherwise, this haplogroup is found in European countries including Italy (including Firenze and Sicily), Belgium, Norway, Germany, and Poland and among Basques from Spain and ethnic French people but also in Adygei (western Circassians) and someone of unspecified ethnicity from Jordan, a country where some Circassians also settled.

One of the Ashkenazic H13a1a1 carriers who tested with Family Tree DNA has a genetic distance of 3 in the Full Coding Region screen from a non-Jewish matriline from sixteenth century Belgium, a non-Jewish matriline from eighteenth century Bavaria, and non-Jewish matrilines from colonial Virginia and early nineteenth century Pennsylvania and all of them are likewise assigned H13a1a1.[229] Another of the Ashkenazic H13a1a1 carriers only sees those non-Jewish matches in the HVR1 screen. Neither of these Ashkenazim match the Iranian Jews as closely as the Full Coding Region screen, only in the HVR1 screen. Therefore, the first Ashkenazic H13a1a1 was probably a Northwestern European who converted rather than a woman of Mizrahi Jewish origin.

Although not closely matching any Ashkenazim, the Iranian Jewish H13a1a1 carriers match three Latin American people with Spanish or Portuguese surnames with a genetic distance of 3 in the Full Coding Region screen.[230] This raises the possibility that those Latin Americans had distant Sephardic Jewish ancestors in their matrilines.

A H13a1a1 carrier from Sardinia was later placed into a new subclade called H13a1a1g by YFull's MTree. Another person from Italy who had been assigned to H13a1a1 was placed into the new subclade H13a1a1j2 by MTree.

H13a1a1 was widespread in ancient southern England.[231] A female carrier of H13a1a1 who died between 300–100 B.C.E. was found in the ancient city of Empúries, which was located in what is today Spain's Catalonia region.[232] A much older H13a1a1 carrier was the male from ancient Russia's Yamnaya culture who died between 3300–2700 B.C.E. and was found at the Ishkinovka I archaeological site in steppelands in the eastern part of Orenburg Oblast, close to Kazakhstan's border.[233]

YFull's MTree estimates that H13a1a1 was formed about 5,000 years ago.

The parent haplogroup, H13a1a, is found in European non-Jews in countries including Norway, Denmark, Poland, and Italy.

H13a1a1a

This is a very rare Ashkenazic lineage that has an overwhelming association with Europeans. It is carried by some people in contemporary Poland[234] and Serbia.[235] It has also been found in numerous Family Tree DNA customers with matrilines from England, Germany, and Sweden as well as some with matrilines from Ireland, Scotland, Norway, Denmark, Czechia, Italy, Spain, Switzerland, Luxembourg, the Netherlands, and Belgium. A particular Ashkenazic carrier of H13a1a1a matches many non-Jewish European H13a1a1a carriers in the Full Coding Region screen in Family Tree DNA, including Christian matrilines from Germany, Denmark, Norway, Sweden, and Finland (but evidently ethnically Swedish) with a genetic distance of 1, matrilines from Germany (in ethnic Germans) and Italy with a genetic distance of 2, and matrilines from England, Scotland, Ireland, and Italy with a genetic distance of 3, among others.[236]

A male carrier of H13a1a1a was buried at the St. Mary Spital cemetery in London, England sometime after the Black Death but before the seventeenth century.[237] A male H13a1a1a carrier dating to between 770–430 B.C.E. was found at Kunda, Lääne-Viru, Estonia.[238]

YFull's MTree estimates that H13a1a1a was formed circa 300 C.E.

H15b

H15b is an uncommon mtDNA lineage for Ashkenazim. It is found, for example, in several documented Ashkenazic matrilines from Hungary. Many non-Jewish people in Denmark have it[239] and its subclade H15b1; only the latter includes the mutation A15715G. H15b is also found in non-Jewish people from Sweden, Italy, Albania, North Macedonia, and Serbia and among Armenians from Turkey and Druze people.[240] H15b2 is also carried by Armenians. H15b1 has also been found in Iran, Germany, and Switzerland. Since this haplogroup family exists in both Europe and West Asia and is estimated to have originated about 8,500 years ago by YFull's MTree, it is difficult to know what kind of ancestor Ashkenazim inherited H15b from.

A person who carried H15b was buried in the "VKH 1201" rural cemetery in Tirup, Horsens, Denmark during the Middle Ages.[241]

H15b and H15b1 were both present in ancient Xinjiang.

H25

A low proportion of Ashkenazim have mtDNA haplogroup H25, including several with documented matrilines from the Russian Empire, including Ukraine and Bessarabia. It has also been found among the Bulgarian Jews,[242] a Sephardic population that admixed to an extent with Ashkenazim. Inside Family Tree DNA, at least some of the Ashkenazic

H25 carriers are an exact match to a Brazilian from the town of Correntes in Pernambuco,[243] a state where some Sephardic Conversos had settled centuries ago. This may suggest that the H25 Ashkenazim actually descend from a Sephardic woman.

I do not have any information about H25's origin and Costa's team did not speculate about it.

H26c

A small proportion of Ashkenazim, including those with matrilines from Belarus, Ukraine, and Moldova, belong to the mtDNA haplogroup H26c. Costa's team found that H26c is one of the Ashkenazic lineages that "nest deep within European lineages in the mitogenome tree."[244] Among modern non-Jews, H26c has been found in Germany, Romania, Finland (in ethnic Finns), the Karelian region of Russia, and the Italian island of Sardinia. The first Ashkenazic H26c was probably a German because the closest match to the Ashkenazim inside Family Tree DNA is a German non-Jew from Dortmund in western Germany, who matches many of the Ashkenazim with a genetic distance of 1 in the Full Coding Region screen.[245]

H26c's parent haplogroup, H26, has been found among non-Jews from Austria, while H26's subclade H26a is found in many modern ethnicities across Europe, including but not limited to non-Jews from Germany, Croatia, Serbia, and Ireland.

In ancient times, H26 was found in three Neolithic era people from what is now Hungary.[246] A male H26 sample dating to between 4500–4000 B.C.E. was a member of the Tiszapolgár population from the Pusztataskony-Ledence archaeological site. Older still is the male H26 sample from the Tisza population at the Vésztő-Mágor site who dates to between 5000–4500 B.C.E. The oldest known H26 sample from Hungary is the female from the Vinča population at the Szederkény-Kukorica-dülö site who dates to between 5321–5081 B.C.E.

H26 was also found in a Neolithic female whose remains, dating to approximately 5500–4850 B.C.E., were found at the Linear Pottery culture's Halberstadt-Sonntagsfeld archaeological site in Halberstadt, Germany.[247] Moreover, one of the victims of the massacre in Potočani, Croatia around 4180 B.C.E. was a female H26.[248]

An old H26a sample was found in Hungary: a male found at the Baden culture's Budakalász-Luppa csárda archaeological site who died between 3340–2945 B.C.E.[249]

H40b

The mtDNA haplogroup H40b has been found in Ashkenazim[250] with matrilines from Poland, Lithuania, Latvia, Belarus, Ukraine, Hungary, and the Netherlands. I do not know whether or not the H40b carrier from Holland who participated in a study by Doron Behar's team is Ashkenazic.[251] H40b does not exist among German Jews so it probably was not found in any medieval Jews in the Rhineland of Germany. YFull places the Holland sample and a Polish Jewish matriline studied by Behar's team into a new branch of their MTree that they unofficially call H40b1.

The Ashkenazic H40b carriers who tested with Family Tree DNA match a fellow H40b carrier who is a non-Jew from east-central Poland. This could point to a Polish origin for this Ashkenazic lineage.

In medieval times, H40b was found in many members of the Kushnarenkovo-Karayakupovo culture in the southern Ural region of Russia. In the Uyelgi cemetery in the Chelyabinsk region, five H40b samples were unearthed from kurgan graves: two dating to the ninth-tenth centuries and three dating to the tenth-eleventh centuries.[252]

Its sister haplogroup H40a is found today in Russia and Denmark and in ancient times in multiple regions, including a Bronze Age male excavated from underneath Yehud, Israel who lived sometime between 2500–2000 B.C.E.[253] Another Bronze Age H40a carrier, dated to between 3200–2900 B.C.E., was found in the Bactria-Margiana Archaeological Complex of Shahr-i Sokhta in far-eastern Iran and sequenced by Vagheesh Narasimhan's genetics team.[254]

Their parent haplogroup, H40, is found today in countries including Portugal and Turkey. A late-medieval child belonging to H40 was found in England.[255] H40 has had a very long-standing presence in Europe, as demonstrated by carriers who were found during archaeological digs in Poland, Hungary, Serbia, and Germany. The H40 from Germany was a member of the Linear Pottery culture who lived sometime between 5500–4800 B.C.E. and found at Viesenhäuser Hof in Stuttgart-Mühlhausen.[256] A Copper Age H40 carrier was found in Veszprém, Hungary dating to between 4339–4237 B.C.E.[257] A Bell Beaker female H40 carrier dating to between 2008–1765 B.C.E. was found in Strachów, Poland.[258] The oldest H40 so far discovered is a male hunter-gatherer from the Lepenski Vir site in Serbia which dates to between 6222–5912 B.C.E.[259]

H41a

H41a has been reliably confirmed to be an Ashkenazic mtDNA haplogroup, including from Romania,[260] Ukraine, and Poland. Ashkenazic members of this haplogroup who tested with Family Tree DNA have a genetic distance of 2 from a few Poles, two Frisians, and an Alsatian in their Full Coding Region screen as of 2021.[261] A variety of other kinds of European non-Jews also have H41a, including British, Swedish, and Danish people and residents of Serbia and Greece. It is also found in Asia, including in China (at least among the Uyghur people in its northwest), in Kyrgyzstan (among ethnic Kyrgyz), and in India.

Two female H41a carriers who died between 880–1000 were buried in the Viking Age mass grave in the churchyard of St. Lars Church in Sigtuna, Sweden.[262]

In Augsburg-Haunstetten, Bavaria, Germany, an archaeological excavation turned up an Early Bronze Age female carrier of H41a whose remains date to between 2033–1918 B.C.E.[263]

The evidence indicates that the first Jewish H41a was a converted European woman, perhaps a Pole.

H47

The mtDNA haplogroup H47 is uncommon among Ashkenazim and appears to originate from either an Italian or an Israelite ancestor though it is also possible the ancestor was Anatolian or Greek. Some Ashkenazic H47 carriers have matrilines from Ukraine and Romania[264] while another traces to the Bavarian region of Germany.[265]

Outside of the Jewish community, H47 has been found in Armenians from eastern Turkey and Armenia and in a Syrian from Damascus as well as in a Palestinian (apparently an Arab) from the West Bank, plus a Maltese person and some modern Italians, including Sicilians and Calabrians. Some of the Sicilians have matrilines from Prizzi, in the

Metropolitan City of Palermo, and from Messina. The Calabrian H47's matriline is from the city of Reggio Calabria, directly across the Strait of Messina from Sicily. Some of these matches were discovered inside Family Tree DNA.[266] Since some Armenians live in Damascus, it is possible that the Syrian match is also of Armenian descent, but the existence of the Palestinian carrier suggests that H47's presence in the Levant is not necessarily due to an Armenian migration. H47 is also present in people in Portugal[267] and Serbia.[268] YFull's MTree contends that this person from Serbia actually belongs to its subclade H47a and also identified H47a carriers from nearby Croatia and Hungary. H47a was already present in Hungary by the eleventh century.[269]

A Bronze Age carrier of H47 who died between 1870–1698 B.C.E. was discovered at the city-state of Alalakh, in what is now Hatay Province in far-southern Turkey.[270]

H47 was the haplogroup of an old man buried at the Vagnari necropolis in southeastern Italy sometime between the first-fourth centuries C.E. when it was part of the Roman Empire.[271] H47 might have migrated into southern Italy from the southern Caucasus or eastern Anatolia. To date, the root level of H47 (that is, H47*) has not been found in modern Italy but branches called H47b and H47b1 do exist there today, with the latter specifically in Sardinia.

For now, I agree with Jonah Stern that Ashkenazim probably acquired H47 from a West Asian ancestor rather than a European one.

H56

A European source is evident for the Ashkenazic mtDNA haplogroup H56 since its Ashkenazic carriers at Family Tree DNA are exact matches in their Full Coding Region screen to an ethnic German from Baden, an Italian speaker from southern Switzerland, an Italian from northern Italy, an Irish person, and a person from the island of Guernsey.[272] H56 is also found among people in modern Sweden and Denmark.

Ashkenazic carriers of H56 include families with matrilines from Ukraine[273] and Lithuania. Doron Behar's team found it in a Sephardic matriline from Spain.[274] It is not clear whether or not the example of H56 in modern Poland discovered by Agnieszka Piotrowska-Nowak's genetics team has any Ashkenazic admixture.[275]

Among its branches, H56a is found among the Irish, H56a1 in Bulgaria and Russia, and H56e in Denmark.

H65a

Eastern Ashkenazim with matrilines from places including Romania, Poland, Ukraine, and Belarus are occasional carriers of the mtDNA haplogroup H65a. This lineage was described as "a recent European haplogroup" by David Collins' team's inquiry into other haplogroups.[276] Available evidence indicates that H65a originated within Europe. An important indicator is the fact that Ashkenazic H65a carriers sequenced by Family Tree DNA match many ethnic Germans. Another is that H65a is also found in non-Jewish matrilines from Ireland and Spain.

H65a's parent haplogroup, H65, has been encountered, albeit infrequently, among modern Poles,[277] Germans, Swedes, and people from Czechia and Switzerland. H65 has also been found in several people from Italy, including a modern Sicilian.[278] There is also an H65 carrier with a matriline from Syria.

HV0-T195C!

The mtDNA haplogroup HV0 (a branch of HV) is rarely found in Ashkenazim but I confirmed the existence of several fully Ashkenazic carriers of HV0 inside Family Tree DNA and 23andMe, including at least one with an Ashkenazic matriline from the Russian Empire. The Family Tree DNA carriers are designated as HV0-T195C! (the exclamation point represents a back mutation).

HV0-T195C! and its branches (including HV0b, HV0c, HV0d, HV0e, HV0f, and HV0g) are also found among non-Jewish people in Switzerland, Italy, Scotland, Ireland, Sweden, Portugal, Greece, and other European countries although HV0c is also found in Palestinian Arabs. HV0 is also common in Mozabite Berbers and Zenata Berbers in North Africa. Modern cases of HV0-T195C! are known from Algeria, Tunisia, the United Arab Emirates, Turkey, Armenia, Portugal, Spain, France, Belgium, Wales, and Northern Ireland, among other countries. Family Tree DNA's Italian participants who carry HV0-T195C! include people with Sicilian matrilines from Graniti and Montalbano Elicona in the province of Messina and from the island of Ustica and one person with a matriline from Melicuccà, Calabria.[279]

Ferragut's team listed "HV0+195" as a haplogroup that is present among the Chueta people of Spain,[280] who largely descend from Sephardic Conversos.

All of the Ashkenazic HV0-T195C! carriers match two people from Spain (at least one likely a Chueta) at a genetic distance of 3 in Family Tree DNA's Full Coding Region screen, including one from Palma de Mallorca where many Chuetas live, but they are slightly genetically closer to certain northern and central Europeans, matching non-Jews from northern Poland, northeastern Germany, Sweden, and England at a genetic distance of 2. The Ashkenazim also have a genetic distance of 3 from one Gulf Arab whose matriline could be Iraqi or Qatari.[281]

Several different branches of HV0 also have Jewish associations. HV0d is found among several Family Tree DNA customers with matrilines from Moroccan Jewish women. HV0b with the mutation 8520G has been found in Portuguese Crypto-Jews from both Bragança and Belmonte and therefore was characterized as "a Sephardic Jewish founding lineage" by Inês Nogueiro's team.[282] However, HV0d is also found among the Basque people,[283] who do not have any Sephardic admixture.

Costa's team's analysis determined that HV0 originated in Europe.[284] Behar's team concluded more specifically that HV0's presence among Jews is due to "Iberian admixture."[285]

Pre-modern carriers of the Ashkenazic sequence HV0-T195C! have been found in both Europe and the Levant. A Neolithic woman who carried HV0-T195C! and died circa 3343–3020 B.C.E. was found at Ballynahatty in Northern Ireland.[286] GenBank lists an HV0-T195C carrier from ancient Lebanon and an HV0-T195C carrier from medieval England, meaning that they carried the T195C mutation, but both are displayed without the exclamation point at the end of their haplogroup designations.[287]

HV1a'b'c

Family Tree DNA confirmed that HV1a'b'c is the terminal mtDNA haplogroup of multiple Ashkenazim, including those with documented matrilines from the Russian Empire. Their close matches inside their Full Coding Region screen are Europeans and Middle Easterners. At least some of them have a genetic distance of 2 from people with matrilines that appear to be French, Polish, and North Germanic or Low German and a genetic distance of 3 from people from Lebanon and Gaza.[288]

A modern carrier of HV1a'b'c from an unspecified part of Italy was found by Giuseppe Gasparre's research team and later included in an analysis by De Fanti's team along with HV1a'b'c carriers from the town of Aviano and the city of Vicenza, both in northern Italy.[289] YFull's MTree recently determined that these samples belong to subclades of HV1a'b'c. Specifically, the Italian from Vicenza is designated as belonging to HV1a'b'c1* by MTree based on the mutation T3906C, while the Italian from Aviano and Gasparre's Italian belong to a daughter branch of that which they call HV1a'b'c1a based on the extra mutation T9018C. Additional Italian HV1a'b'c carriers tested with Family Tree DNA, including one with a matriline from Campobasso in southern Italy and another with a matriline from Casoni di Mussolente in northern Italy.

Other modern people whose haplogroup is listed as HV1a'b'c in GenBank include an Iranian, an Armenian, and two Palestinian Arabs.[290]

Ancient cases of HV1a'b'c are known from southeastern Europe, Egypt, and the Levant. A young child who belonged to HV1a'b'c was buried in a mass grave at Shekerdja Mogila near the village of Kamen in east-central Bulgaria dating to the Early Bronze Age.[291] Three HV1a'b'c carriers were found in tombs at the site of the ancient city of Abusir el-Meleq in Egypt: a male from the Roman era who was buried sometime between the years 26 and 84 C.E. and two people from the Ptolemaic period who were buried sometime between the fourth century B.C.E. and the second century B.C.E.[292] A HV1a'b'c carrier was found in an Iron Age necropolis at the archaeological site of Polizzello underneath today's town of Mussomeli in Sicily.[293]

A much earlier HV1a'b'c sample is the male who was buried sometime between 4500–3900 B.C.E., during the Chalcolithic era (Copper Age), in Peqi'in Cave in Upper Galilee, Israel.[294] This sample suggests that Ashkenazim could have inherited this haplogroup from an Israelite ancestor.

HV1b2

HV1b2, a descendant haplogroup of HV1a'b'c, is the fifth most common mtDNA haplogroup among Ashkenazim. It is found in French

Jewish, German Jewish, Austrian Jewish, and Czech Jewish matrilines as well as Eastern Ashkenazic ones from places that are now in Poland, Lithuania, Latvia, Belarus, Ukraine, Slovakia, Moldova, Romania, and Hungary.[295] Genetic evidence supports HV1b2's potential Israelite origin for Ashkenazim but it could have also derived from another part of West Asia.

HV1b2 has been found in members of multiple other Jewish communities, including in a Family Tree DNA tester in the "Romaniote DNA Project" with a matriline from the Romaniote Jewish community from Larissa, Greece[296] and, as Eugene Dubrovitzky discovered, in multiple members of the Mountain Jewish community from Quba, Azerbaijan. Outside of the Jewish world, HV1b2 has also been found in an Armenian with a matriline from the city of Muş in eastern Turkey who participated in a study by Miroslava Derenko,[297] a Yezidi Kurd from Georgia,[298] and an Emirati from Ras Al Khaimah.[299] A Sicilian from Messina in northeastern Sicily tested at Family Tree DNA and got assigned to HV1b2.[300]

An Egyptian mummy from the Ptolemaic period that dates to sometime between 394–239 B.C.E. and was found in a tomb at the site of the ancient city of Abusir el-Meleq was assigned to haplogroup HV1b2.[301]

Costa's team of geneticists wrote that HV1b2 "clearly nests within an Armenian/Anatolian subcluster of the Near Eastern clade HV1b."[302] Six years later, Shamoon-Pour's team wrote that HV1b2 has "links to Northern Mesopotamia" and explained as follows:

> Our updated phylogeography of HV1b-152 suggests its pre-Neolithic origin in South Caucasus or northern Mesopotamia, with HV1b2 positioned close to two Assyrian branches. Given our age estimates for HV1b2, it is conceivable that this clade originated among the displaced Jewish communities during Assyrian and Babylonian captivities (2.5–2.7 kya), and likely remained exclusive to the Jewish populations that settled in Upper Mesopotamia, including Adiabene and Osroene Kingdoms, up to 1.7 kya.[303]

Some modern Assyrians and Italians carry HV1b2's parent haplogroup, HV1b. Other branches of HV1b similarly have a presence in the Mediterranean region and the Middle East. The variety called

HV1b-T152C is found in Assyrians, Armenians, and Tunisians. Other Armenian haplogroups include HV1b4, HV1b5, HV1b3a, and HV1b3b, with HV1b3b also present in Iran. HV1b1a and HV1b1b are both found in Yemen and Somalia. HV1b6 is found in Italy and HV1b3 is found in Sardinia. YFull's MTree designates the parent haplogroup of both HV1b2 and HV1b3 as HV1b-a and defines that by the presence of the back mutation T152C!.

Ancient members of the HV1b family were found at a couple of archaeological sites in the Levant. An Amorite member of haplogroup HV1b3b dating back to sometime between 1497–1326 B.C.E. was found at the city-state of Alalakh, in what is now Hatay Province in far-southern Turkey.[304] Amorites spoke a Northwest Semitic language and inhabited Syria and Israel before the Israelites. A Canaanite assigned to haplogroup HV1b3 who died sometime between 1800–1700 B.C.E. was found at Tel Hazor in Israel's Upper Galilee region.[305]

Musilová's team determined that the ancestral haplogroup HV1 first "appeared . . . in the Near East."[306]

HV5

I received confirmation, and later confirmed myself, that HV5 (also called HV5*) is the completely sequenced mtDNA haplogroup assignment for several full Ashkenazim who tested at Family Tree DNA. At least some of their matrilines trace back to Poland and the Russian Empire.

Family Tree DNA's mtDNA Haplotree also lists an HV5 carrier whose matriline comes from Switzerland. This person is possibly not Jewish.

HV5a

HV5's subclade HV5a is much more common in Ashkenazim than is HV5. HV5a is encountered in Ashkenazic matrilines[307] from Poland, Ukraine, Belarus, Lithuania, Latvia, Hungary, Romania, Moldova, Slovakia, Austria, Switzerland, Germany, France, and the Netherlands.

Elsewhere, HV5a is also found among Armenian people[308] and people in India.[309] Its presence in Armenians suggests that Ashkenazim may have inherited this from a West Asian ancestor.

Its sister haplogroup HV5b is found among Italians, Iranians, and Pakistanis.

I1c1a

The mtDNA haplogroup I1c1a has been confirmed to exist in Ashkenazic matrilines from Germany, Austria, Hungary, Romania, Moldova, Ukraine, Poland, and Lithuania.[310] Behar's team found a I1c1a carrier with a matriline from Turkey[311] and there is a Chechen carrier of I1c1a who tested with Family Tree DNA and two Hungarian I1c1a carriers who uploaded their mtDNA to YFull's MTree.

I1c1a's parent haplogroup, I1c1 (also called I1c1*), has been found in various non-Jewish European and Middle Eastern populations. An Italian I1c1* matriline tested at Family Tree DNA and a Northern Italian I1c1* case was studied by Anna Olivieri's team. I1c1* carriers have also been studied whose matrilines trace back to Poland, Ukraine, Bashkortostan, and the United Arab Emirates.

Olivieri's team established that an ancient person found at the Neolithic necropolis of Camí de Can Grau in Spain dating back to between 3500–3000 B.C.E. belonged to haplogroup I1c1.[312] While acknowledging I1c1's presence in southern Europe, including in this sample from Camí de Can Grau, Costa's team was inclined to believe that it "may have originated in the Near East."[313] Olivieri's team thought that "I1c1 was likely a marker of Neolithic dispersal in Europe" and that the Ashkenazic carriers of I1c1a did not inherit it from a more recent Middle Eastern ancestor such as an Israelite.[314] That being said, a case of I1c1 is known from the ancient Levant: in the early Roman era, sometime between 176 B.C.E. and 3 C.E., a male was buried at a cemetery unearthed in Beirut, Lebanon who was a carrier of this maternal haplogroup,[315] but not of the Ashkenazic subclade I1c1a. A male steppe nomad who carried I1c1 and died around 300 C.E. was found at a Otrar-Karatau cultural site in Kazakhstan.[316]

YFull's MTree identified an intermediate haplogroup descended from I1c and ancestral to I1c1 that they call I1c-a and they found it in a modern YFull tester with a matriline from Barcelona, Spain. They also assign the aforementioned ancient sample from Beirut to I1c-a. I1c-a carriers have the mutation C16270T but do not have the mutation T9386C that defines I1c1.

I1c1a's even deeper ancestral haplogroup, I1c, has been found in people from Spain, Germany, Turkey, and Kuwait. An I1c carrier with a matriline from Turkey was confirmed to be an ethnic Turk by Ian Logan.

I5a1b

The mtDNA haplogroup I5a1b (also called I5a1b*) is exceptionally rare among Ashkenazim. It is the confirmed haplogroup assignment for a Family Tree DNA customer with a Ukrainian Jewish matrilineal ancestor who had a distinctly Ashkenazic maiden name. I also became aware of a person with a fully Ashkenazic autosomal DNA profile and recent family history, including a direct maternal line from Belarus, whose haplogroup was partially sequenced as I5a by 23andMe. This person must certainly belong to I5a1b because that is the only variety of I5a that Ashkenazim ever possess.

I5a1b* is also found in some people today in Italy and Spain. Reginaldo Ramos de Lima, a Brazilian carrier of I5a1b, expressed his belief that he inherited this haplogroup from a Sephardic Crypto-Jewish ancestor.[317] A branch of I5a1b that YFull's MTree calls I5a1b1, with C494A as the defining mutation, was found in a Brazilian matriline from Bahia and a Brazilian matriline from Piauí as well as a person with a matriline from Bragança, a city in northeastern Portugal where many of the inhabitants have Sephardic Crypto-Jewish heritage. It is therefore possible that the Ashkenazim with I5a1b had a Sephardic matrilineal ancestor.

The ancestral haplogroup I5a1 is found among Assyrians and in Bulgaria, Serbia, Montenegro, Germany, England, Ireland, Poland, and Finland.

Two Syrian Jews who tested at 23andMe are assigned to haplogroup I5a1 but that is not necessarily their terminal haplogroup because that company does not fully sequence all haplogroups.

J1b1a1

The mtDNA haplogroup J1b1a1 is found among Ashkenazim with matrilines from Germany, Austria, and Poland. J1b1a1 is also found widely in non-Jews across Europe, including in England, Scotland, Denmark,

Sweden, Belarus, Poland, Croatia, France, Spain, Italy (among Sicilians), and other countries. In West Asia, J1b1a1 is found among Armenians from Turkey and among Assyrians. Elsewhere in Asia, J1b1a1 has a presence in India, Kazakhstan, the Altai region, and the Pamir region.

Cases of J1b1a1 are known from medieval and ancient Europe and Asia, too. For instance, a Late Bronze Age grave unearthed at the Ķivutkalns cemetery in Latvia contained a male who belonged to J1b1a1 and died between 405–230 B.C.E.[318] A J1b1a1-carrying female from the Corded Ware culture who died circa 2863–2498 B.C.E. was found at Tiefbrunn in Germany.[319] A Copper Age site in Augsburg, Bavaria, Germany from the Bell Beaker culture contained two female carriers of J1b1a1 who died circa 2500–2000 B.C.E.[320] An Iron Age nomadic male found at the Karagaly 1 burial site in Kazakhstan's Tian Shan region also carried J1b1a1 and he died between 790–747 B.C.E.[321]

Ashkenazic J1b1a1 carriers have the extra mutation C2322T which YFull's MTree uses to define the novel subclade J1b1a1f that includes a YFull customer with a matriline of unlisted ethnicity from Ukraine.

Other versions of J1b1a1 with extra mutations are found today in many countries including Poland, Hungary, Serbia, Tunisia, and Armenia and one of them was in Roman-era Italy, Iron Age Armenia, and Iron Age Kazakhstan. Daughter subclades of J1b1a1 have also been encountered in Europe such as J1b1a1a and J1b1a1e in pre-modern England, J1b1a1d in Italy and among medieval Langobards in Hungary, and J1b1a1b and J1b1a1g in Denmark. J1b1a1e also exists outside of Europe, specifically in modern Iran.

Its sister haplogroup J1b1a3 has been found in a Syrian Jewish matriline from Aleppo.

J1c-C16261T

Carriers of the mtDNA haplogroup J1c are scattered across Europe and the Middle East, ranging from modern Bulgarians to Assyrians and from ancient peoples from Sicily to Germany. Costa's team cited the assessment that the first J1c lived in "Late Glacial Europe."[322] Most J1c carriers have the mutation G185A and G228A. The Ashkenazic variety of the base haplogroup J1c contains the extra mutation C16261T, which is why Family Tree DNA and GenBank call it J1c-C16261T. The Ashkenazim also carry the mutations T16093C and C16519T and the back mutation G16274A!. This is found infrequently in Ashkenazim with matrilines tracing back to Poland and Lithuania. At least one Ashkenazi with a matriline from Poland had a complete mitochondrial sequence done.

Maria Pala's team genetically sequenced four carriers of J1c-C16261T from Greece, Romania, Bosnia, Armenia, and Kuwait.[323] Several Bulgarian customers of Family Tree DNA also carry J1c-C16261T. Other Family Tree DNA testers with J1c-C16261T include a Sicilian from Sciacca (a town in the province of Agrigento) and mainland Italians with matrilines from places including but not limited to Tolve (a town in the province of Potenza in southern Italy), Avellino (a town in the Campania region of southern Italy), and Caulonia (a municipality in the Calabria region of southern Italy).[324] Boris Malyarchuk's team, meanwhile, located a J1c-C16261T carrier in modern Poland[325] whom I presume was not Jewish.

A female from the Middle Helladic culture whose haplogroup is listed as "J1c+16261" (equivalent to J1c-C16261T) by Clemente's team was buried in a pit grave at the archaeological site of Elati-Logkas in northern Greece around 2000 B.C.E.[326] Interestingly, a "J1c+16261" (J1c-C16261T) carrier who was apparently a Phoenician lived in ancient Lebanon.[327] So it is unclear whether Ashkenazim inherited this haplogroup from an Israelite, a Greek, an Italian, or a Pole.

J1c1

The mtDNA haplogroup J1c1 has been sequenced in Ashkenazim with matrilines from Belarus and Saint Petersburg, Russia. It is also found among non-Jews in numerous European countries including England, Scotland, Norway, Sweden, France, Luxembourg, Italy (including Sardinia and Sicily), and Slovenia. Some of the Sicilian carriers descend from a woman from the village of Cattolica Eraclea in the province of Agrigento.[328] The two J1c1 samples in GenBank from people with matrilines from Poland, including one collected by Piotrowska-Nowak's team from a contemporary person living in Poland, are potentially non-Ashkenazic.[329] J1c1 is also occasionally found outside of Europe, such as in a Palestinian Arab[330] and supposedly also in Uzbekistan.

J1c1 samples from medieval and ancient Europe are abundant. J1c1 existed in medieval Denmark.[331] Earlier J1c1 samples come from the

Neolithic Age, Copper Age, Bronze Age, and Iron Age. A Neolithic J1c1-carrying male who died circa 3700–3639 B.C.E. was found in Distillery Cave in Oban, Scotland.[332] J1c1 was found in three females from the Fatyanovo culture in Russia from the 3rd millennium B.C.E.[333]

Spain is a particular hotspot for ancient J1c1 samples. At the Camino del Molino burial site in Caravaca, Murcia, Spain a male carrier of J1c1 was laid to rest circa 2920–2340 B.C.E.[334] In El Mirador Cave in Atapuerca, Burgos, Spain a female carrier of J1c1 was found dating from circa 3363–1903 B.C.E.[335] A male J1c1 carrier was found at Cova de la Guineu, Font-rubí, Barcelona, Catalonia in eastern Spain who had died circa 3400–2500 B.C.E.[336] A male J1c1 who died between 3500–2900 B.C.E. was found at Mandubi Zelaia, Ezkio-Itsaso, Gipuzkoa in Spain's Basque country.[337] A J1c1 male who died circa 3090–2894 B.C.E. was found at La Chabola de la Hechicera in Alava in Spain's Basque country and a J1c1 female was found at the same site dating from circa 3014–2891 B.C.E.[338] A J1c1 male who died circa 515–375 B.C.E. was found at Mas d'en Boixos-1 in Pacs del Penedès, Barcelona, Catalonia, Spain.[339]

Even older J1c1 carriers were found in the Balkans and central Europe. At Paliambela in northern Greece, researchers found a female J1c1 who had died between 4452–4350 B.C.E.[340] In the village of Govrlevo in what is now North Macedonia, a J1c1 male was found whose remains were dated to between 5979–5735 B.C.E.[341] A male member of the Linear Pottery culture in the Pannonia region, today Hungary, was found at Kompolt-Kigyoser and dated to between 5210–4990 B.C.E.[342]

There are numerous branches of J1c1 and they are found across modern Europe. As an incomplete list, J1c1a is found among Finns and Danes, J1c1b is found in Irish and Basque people, J1c1b1 is found in Portuguese and English people, and J1c1d is found in Danes. J1c1d is attested from medieval England while J1c1b is attested from medieval Norse in Greenland and from ancient Switzerland, Germany, and Poland.

J1c3e2

J1c3e2 is an uncommon mtDNA haplogroup among Ashkenazim. It has been confirmed in Ashkenazic matrilines from Poland, Lithuania, Belarus, and Ukraine. J1c3e2 also was found in a non-Jewish Northern

Italian matriline, a Sardinian matriline,[343] a matriline from an unspecified region of Italy,[344] and matrilines of unknown ethnicities from Spain, Denmark, and Sweden. YFull's MTree places the Denmark sample into its newly named subclade J1c3e2a and the Sweden sample into its newly named subclade J1c3e2b.

A J1c3e2-carrying male who was either a gladiator or a soldier was buried at Driffield Terrace in England's Yorkshire region during the Roman era.[345] Isotomic analysis of his remains confirmed that he was not originally from Britain but instead probably had lived in the Mediterranean region, either southern Europe or North Africa.[346] A pre-Roman J1c3e2 carrier was found at the Colfiorito necropolis in Italy's Umbria region.[347]

The first Jewish J1c3e2 woman was probably a European convert, likely an Italian.

J1c4

J1c4 (also called J1c4*) is a very rare mtDNA haplogroup for Ashkenazim. In Family Tree DNA and GEDmatch, I confirmed that J1c4 is assigned to at least two full Ashkenazim. One of them has documentation tracing their matriline to an Ashkenazic woman from Bavaria, Germany. Another has an Ashkenazic matriline from Hungary or Romania.

J1c4 has a broad distribution in Europe's non-Jewish ethnicities. It is especially prominent in ethnic Germans and also in people from England, Sweden, Poland, and Scotland if we judge by Family Tree DNA's mtDNA Haplotree. People from Spain in the southwest to Finland in the northeast have been sequenced as J1c4. It is also found in Belgium's French-speaking population and in Denmark, Norway, Switzerland, Luxembourg, France, Wales, Italy, Serbia, and Estonia, among other countries. Some modern people in Poland who are presumably not Ashkenazim also carry J1c4. YFull's MTree found this haplogroup in a Turk from Turkey.

Descendant haplogroups of J1c4 are likewise mostly found in Europeans. J1c4b is found in non-Jews in Russia, Ukraine, Poland, Bulgaria, Serbia, Croatia, Austria, and Germany but also in Uyghurs and people in Lebanon and Georgia. Its daughter J1c4b1 is also found in Russia. J1c4c is in Ireland, France, and the Netherlands. J1c4d is in

Russia, Bulgaria, Poland, and Lithuania. J1c4f and J1c4g are in Italy, with the former attested in its regions of Sardinia and Umbria.

Since J1c4 exists in German Jews and also in many German non-Jews, my impression is that Ashkenazim probably acquired this haplogroup from a German convert to Judaism.

J1c5

The mtDNA haplogroup J1c5 (also called J1c5*) is rarely found among Ashkenazim. It was detected in the Ashkenazic population by Costa's team.[348] An Ashkenazic carrier of J1c5 with a matriline from Ukraine was cited by Pala's team.[349] Fully Ashkenazic carriers of J1c5 also tested with Family Tree DNA and 23andMe. A J1c5 carrier at Family Tree DNA has an Ashkenazic matriline from Hungary and in the Full Coding Region screen matches a Methodist Christian with a matriline from seventeenth-century England at a genetic distance of 3.

J1c5* is also found in non-Jewish Europeans including Poles, Russians, Finns, Swedes, English, Scots, and Italians, among others, and is also in Algeria, Kazakhstan, Siberia, Yakutia (among the Yakuts), and India.

J1c5* carriers are known from pre-modern Europe. The haplogroup was present in Hungary by the eleventh century. A Bell Beaker female with this haplogroup was found at Quedlinburg's Site IX in Germany and she died circa 2346–2033 B.C.E. during the Copper Age.[350] Other J1c5 carriers from Germany were two males from the Corded Ware culture who were laid to rest in Esperstedt between 2500–2050 B.C.E.[351] A Neolithic female from the Linear Pottery culture in Pannonia (today's Hungary) dating from 5211–5011 B.C.E. was also in J1c5 and was found at Polgár-Ferenci-hat.[352] A Neolithic J1c5 female was placed in a grave at Carcea, Romania circa 5484–5372 B.C.E.[353] Two males from the Funnel Beaker culture who were found in the early Neolithic era megalithic cave in Ansarve in Tofta Parish on the Swedish island of Gotland belonged to J1c5.[354]

Subclades of J1c5 are similarly common in Europe. To provide some examples, J1c5a is found in non-Jews in countries like Russia, Latvia, Sweden, Poland, Scotland, and Ireland and on Spain's Canary Islands, J1c5c is found in Greece and Kosovo, and J1c5f is found in Ireland and was in Bronze Age England.

J1c5 almost certainly arrived in the Ashkenazic population through a European woman who converted to Judaism.

J1c7a

The mtDNA haplogroup J1c7a had a very wide geographical distribution among Ashkenazim in Europe. It has been found in people with Ashkenazic matrilines[355] from Germany, Austria, Switzerland, France, Belgium, Czechia, Slovakia, Hungary, Romania, Lithuania, Latvia, Ukraine, Belarus, and Poland. This had been called J1c7a1a in Costa's team's study of Ashkenazi mtDNA[356] but that is not its current haplogroup assignment at Family Tree DNA or YFull's MTree. Outside of the Ashkenazic community, J1c7a has a presence among modern Poles, Germans, Swedes in Sweden and Finland, Finns, and the English, among other non-Jewish Europeans, plus Persians from the province of Razavi Khorasan in northeastern Iran. Some Swedes and Finns also belong to J1c7a's branch J1c7a1.

A medieval Hungarian sample from Karos-Eperjesszög cemetery number three who died between 890–950 was a carrier of J1c7a.[357] Among the remains found at the Wielbark culture's cemetery at Kowalewko in Greater Poland Province in today's west-central Poland that dates to between about 50–220 C.E., during the region's Roman Iron Age, was a female carrying J1c7a.[358] Archaeological evidence suggests that this community descended from members of the Oksywie culture of Poland but that they had possibly admixed with some Gothic immigrants who arrived from Scandinavia. A male sample from the eastern Swedish island of Gotland in the Baltic Sea that dates back to approximately 1610–1440 B.C.E., during the Early Bronze Age, was also identified as a J1c7a carrier.[359]

The most likely source of J1c7a for Ashkenazim would be a German rather than a Slavic person like a Pole, even though its introgression from an eastern European person was suggested by Marta Costa's genetic research team.[360] One reason is because German Jews have it. Another is because there are far fewer Slavic matches than Western European matches to this lineage.[361] Leo Cooper wonders whether the first Jewish J1c7a lived near Germany's Elbe River.

J1c7a's ancestral root, J1c7, is present among non-Jews from European countries including Finland, Austria, Slovakia, Romania, Bosnia and Herzegovina, Italy, and Greece. A 2012 study by Pala's team had gathered two European samples (one from Spain, another from an ethnic Belarusian) that they listed as J1c7a1 and submitted to GenBank with that assignment[362] but YFull's MTree places both samples into the root level haplogroup J1c7* instead.

A daughter subclade called J1c7c is found among the Kashkuli people, one of Iran's Qashqai tribes.[363]

J1c13

Among Ashkenazim, J1c13 is considerably less common than are J1c1, J1c7a, and J1c14. J1c13 exists in Ashkenazim with matrilines from Ukraine, Poland, and Austria. An Italian from the province of Salerno in southern Italy, a Romanian, and a Serbian who all had their mtDNA sequenced by Family Tree DNA show as matches to the Ashkenazim. J1c13 is also assigned to some people from Greece, Turkey, Germany, and Sweden who tested with that company.

J1c14

So far, Ashkenazic J1c14 carriers[364] only match other Ashkenazim. It is therefore not possible to say much about this lineage. These Ashkenazic matrilines trace back to Ukraine, Poland, Lithuania, Latvia, Belarus, Romania, Slovakia, Hungary, Austria, and Germany. J1c14 was presumably carried by some early medieval German Jews as well. J1c14 was previously called J1c7d.

J2b1e

The mtDNA haplogroup J2b1e is found in some Ashkenazim with matrilines from Ukraine, including but not limited to the formerly Hungarian region in far-western Ukraine, as well as from Hungary, Poland, and Lithuania. Inside Family Tree DNA, I found a Sephardic Jewish J2b1e carrier whose matriline came from Livorno, Italy.

Elsewhere, J2b1e has been found by professional geneticists among non-Jews in Lebanon and Italy.[365] A member of Family Tree DNA's Cyprus project whose matriline is Cypriot Greek also belongs to J2b1e.[366]

It has a sister haplogroup called J2b1f that is found among non-Jews from Lebanon, Syria, Turkey, and the United Arab Emirates. In addition, Mountain Jews from Azerbaijan and Daghestan often carry J2b1f.[367] A branch called J2b1f1 exists among Armenians.

The parent haplogroup of J2b1e, called J2b1, is found among Italians (including Sicilians and Umbrians) as well as people in Morocco, Egypt, and Syria, but also in Germany, England, Poland, Finland, and Russia.

It is likely that Ashkenazim acquired J2b1e from an Israelite ancestor.

K1a1b1

The mtDNA haplogroup K1a1b1, also called K1a1b1*, came to the Ashkenazic population from a woman who was different than the woman (or women) who brought the much more prevalent subclade K1a1b1a. The Ashkenazic variety of K1a1b1* does not contain the HVR1-level mutation C16234T that K1a1b1a carriers have, and Ashkenazic K1a1b1* carriers are also missing the HVR1-level mutation T16223C that most (but not all) K1a1b1a carriers have. However, they do share the mutations C114T, A10978G, and T12954C with K1a1b1a. K1a1b1* is surprisingly rare among Ashkenazim. At least one of them has a documented Ashkenazic matriline from Poland.[368]

This haplogroup has had a noteworthy presence in western Europe in ancient and modern times. Modern carriers of K1a1b1* include members of the Basque, Portuguese, English, French, Belgian, German, Danish, and Swedish peoples, all of them non-Jewish. In Family Tree DNA, within the Full Coding Region screen, the Ashkenazic K1a1b1* carriers have a genetic distance of 3 from non-Jewish K1a1b1* carriers from Scandinavia, Ireland, Germany, and Spain.[369] This reveals that the first Ashkenazi K1a1b1* would have been a European person. K1a1b1a carriers do not match them in that screen.

K1a1b1* was also in central Europe, specifically in an old adult female in medieval Hungary.[370] Her sequence was confirmed by YFull's MTree.

K1a1b1's most ancient samples were found in Spain, Morocco, and Great Britain. A Neolithic male K1a1b1 carrier was buried sometime between 3960–3710 B.C.E. at the archaeological site Les Llometes in Valencia in eastern Spain.[371] Another Neolithic male K1a1b1, who died between 3900–3600 B.C.E., was found at the site La Mina in Spain.[372] During the subsequent Copper Age, K1a1b1 was commonplace among members of the Bell Beaker culture in Spain. A Bell Beaker male and a Bell Beaker female who both belonged to this haplogroup were buried at the archaeological site Arroyal I in Burgos, Spain between the 2450s-2200s B.C.E.[373] Two Bell Beaker male K1a1b1 carriers from around 2500–2000 B.C.E. were found at the archaeological site Humanejos in Parla, Madrid, Spain.[374] A female K1a1b1 who died circa 2200–2000 B.C.E. was found at the megalithic burial site La Navilla in Arenas del Rey in Spain's Granada province.[375] The haplogroup continued to be present in Spain into the Bronze Age, as demonstrated by its female carrier from the Cogotas I culture who died between 1368–1211 B.C.E. and was found at the site La Requejada in San Román de Hornija in the province of Valladolid.[376] A K1a1b1 carrier from a pre-Phoenician culture who died sometime between 900–750 B.C.E. was found at the megalithic gravesite Ca na Costa on Formentera, one of Spain's Balearic islands.[377]

Two Neolithic K1a1b1 carriers were buried at the Kelif el Boroud archaeological site in Morocco, which dates to between 3780–3650 B.C.E.[378]

A Neolithic male K1a1b1 who died between 4000–3300 B.C.E. was found in Upper Swell, England.[379] A Neolithic female carrier who died between 3360–3098 B.C.E. was found on an island in Orkney, Scotland.[380] A Bronze Age female K1a1b1 who died between 1619–1457 B.C.E. was found in South Lincolnshire in England.[381] A female K1a1b1 who died circa 1608–1502 B.C.E. was found in the niche Grottina dei Covoloni del Broion in northeastern Italy.[382]

YFull developed a new nomenclature for their MTree whereby K1a1b1 and K1a1b1a are separated by an intermediate haplogroup they call K1a1b1-b which includes the ancient sample from Ca na Costa and the two ancient samples from Kelif el Boroud who all had the C114T mutation but not A10978G or T12954C. MTree lists the carriers of all three mutations as K1a1b1a, just as Ian Logan does. Between K1a1b1-b and K1a1b1a is another intermediate haplogroup that MTree calls

K1a1b1-b1 and it is defined by C16234T. Two modern samples that MTree identifies as K1a1b1-b1a carriers were called K1a1b1 in the original study by Neus Font-Porterias's team, which noted that they carry the mutations C114T and A7158G.[383] A7158G is the mutation that MTree uses to define the haplogroup K1a1b1-b1a, which is a daughter subclade of K1a1b1-b1. Both of those samples are Tunisian Berbers.[384]

A young Iron Age Gaul child from the Urville-Nacqueville necropolis in northwestern France who died in roughly 182 B.C.E. was partially sequenced into haplogroup K but later fully sequenced as K1a1b1a by Fischer's genetics team.[385] It certainly has the mutation C16234T, so we know it is at least at the level of K1a1b1-b1.

A particular Ashkenazic person who was sampled by Behar's team was initially classified as belonging to the haplogroup K1a1b[386] but YFull's MTree's new nomenclature classifies it more precisely as a carrier of K1a1b1-b1.

Daughter subclades of K1a1b1 were and are found primarily in Europe. K1a1b1e is found in Scots, Irish, and Tuscans in Italy and was in Neolithic Poland but also Bronze Age Armenia. K1a1b1b, K1a1b1d, and K1a1b1f are all in England and Ireland. K1a1b1b1 is in Sweden and Finland. K1a1b1c was in medieval England and is in modern England and Austria and in Canarians from Spain's Canary Islands. K1a1b1g was in Bronze Age Germany and England and is in modern Greeks.

K1a1b1a

K1a1b1a is by far the most common maternal lineage among Ashkenazim. Many German Jews possess it, so it was almost certainly present among the medieval Jews of the Rhineland of Germany. It is also found among many Polish Jews, Ukrainian Jews, and Belarusian Jews as well as among Lithuanian Jews, Latvian Jews, Estonian Jews, Hungarian Jews, Czech Jews, Slovakian Jews, Romanian Jews, Croatian Jews, Austrian Jews, Swiss Jews, French Jews, and Dutch Jews. Costa's team found it at a collective rate of 37.5 percent among Jews from Germany, Switzerland, and the Netherlands.[387]

Voluntary conversions of thousands of Ashkenazic Jews to Roman Catholicism in L'viv (then Lwów) and elsewhere in Poland, particularly in the 1750s and 1760s, were responsible for introducing K1a1b1a into the ethnic Polish population. Most, but not all, of these converts had belonged to the Frankist sect.[388] It is known that the bulk of Poles with distant Ashkenazic ancestry lived in the nineteenth and twentieth centuries in southeastern Poland and southwestern Ukraine, but K1a1b1a has even been found in Poles from Upper Silesia and the Gdańsk region in addition to some in southeastern Poland.[389] Another result of assimilation was that there are K1a1b1a descendants in the Polska Roma community in western Poland.[390]

This haplogroup is also found in other Jewish and Jewish-descended populations. K1a1b1a has been confirmed among Sephardic Jews from Bulgaria, North Macedonia, and Turkey.[391] It is possible that the haplogroup arrived in those communities as a result of their minor Ashkenazic admixture but that is not necessarily the explanation for most of them. It does appear that K1a1b1a would have been present among the Sephardic Jews in medieval Spain because it has been found in a low percentage of Chuetas, descendants of Sephardic Conversos who live in Mallorca, Spain.[392] The Chuetas do not descend at all from Ashkenazim. K1a1b1a has also been found in a Jewish family from India whose ancestors were Iraqi Jews from Baghdad. Additionally, K1a1b1a was found in four members of the Jewish community of Cochin, India.[393]

Some researchers proposed that the first Jewish K1a1b1a was Middle Eastern while others proposed a European ancestor. Costa's team suggested that its origin may be in southern or western Europe.[394] This remains possible for the Sephardic and Ashkenazic groups. It makes less sense for the founder of K1a1b1a in Cochin Jews unless Gyaneshwer Chaubey's team is correct that it could have been "transmitted through Paradesi Jewish" people (descendants of Sephardic Jews who arrived in Cochin well after the Cochin Jews).[395] Otherwise, as Chaubey's team wrote, the explanation for the haplogroup's presence in Cochin Jews would be a woman from the "Middle East during the initial settlement."

Family Tree DNA's mtDNA Haplotree lists K1a1b1a carriers with matrilines from Syria, Azerbaijan, and Uzbekistan from unspecified ethnicities. This haplogroup also has a surprising presence in one group of non-Jews in South Asia: 5 percent of members of the Khattak tribe of Pashtuns in northwestern Pakistan who participated in a scientific study by Shahzad Bhatti's team were sequenced as K1a1b1a.[396] Autosomal DNA evidence does not support the myth that Pashtuns are one of the lost tribes of Israel.[397] Bhatti's team claimed that the presence of K1a1b1a in the Khattak "was a result of an ancient genetic influx in the early Neolithic period." That would have been far before Judaism was founded. That claim is contradicted by YFull's MTree, which shows that the haplogroup originated more recently: about 4,000 years before present generations, that is, around 2050 B.C.E.

K1a4a

Among confirmed Ashkenazic K lineages, K1a4a (also called K1a4a*) has the second-lowest frequency among Ashkenazim. It is found among German Jews and Ukrainian Jews.

It may be significant that K1a4a is also the haplogroup assignment of a Sicilian from Cefalù, a city within the Metropolitan City of Palermo, and a Sicilian from Termini Imerese, a town that is also within Palermo, and of multiple Greeks, including one with a matriline from the town of Astros in the northeastern Peloponnese region of Greece and another one with a matriline from Asia Minor, all of whom tested with Family Tree DNA. The Sicilian and Greek participants match one another more closely (on the HVR2 screen) than they match the Ashkenazim (only on the HVR1 screen).[398] There is also a Turkish person matching a Sicilian participant on the HVR2 screen. That Sicilian also matches a Tunisian Jew and a person with a Sephardic Jewish matriline from Egypt on the HVR2 screen and the specific haplogroup K1a4a displays for both of them. It is possible that Ashkenazim inherited K1a4a from an ancient ancestor from Greece or Italy who converted to Judaism.

Other K1a4a carriers in Family Tree DNA's mtDNA Haplotree report having matrilines from Portugal, Spain, France, Ireland, Austria, Romania, and Syria. The presence of K1a4a in the Levant shows that a Levantine originator for this haplogroup in Ashkenazim is not impossible. Scientific studies found K1a4a in Italians from the Umbria region,[399] Sardinians, Basques, and Canarians from Spain's Canary Islands.

Several pre-modern K1a4a samples add to our knowledge. One of the early medieval Muslim Berber graves in the city of Nimes in southern France contained a young man belonging to K1a4a[400] whose remains were dated to between the years 637–765. A K1a4a female died sometime between the first-fourth centuries, probably around the year 200, in the

Roman Empire and was buried in the pagan Isola Sacra Necropolis in what is today the Lazio region of central Italy.[401] Leo Cooper informed me that analysis using Eurogenes G25 showed that this female was autosomally very similar to modern Greeks from the Dodecanese Islands.[402]

A subclade of K1a4a that is called K1a4a1 is found in modern Sicilians from central Italy as well as in non-Jews from other European countries including Switzerland, Luxembourg, Germany, Portugal, Scotland, Ireland, Norway, and Serbia. K1a4a1 is also found among the Basques and the inhabitants of Spain's Balearic island of Ibiza. Outside of Europe, K1a4a1 is present in Morocco and Tunisia.

K1a4a1 is also attested in huge numbers of medieval and ancient samples from Europe and North Africa. For example, a Bell Beaker female who died between 2455–2200 B.C.E. and was found at Amesbury Down in England carried this subclade.[403] Two Bell Beakers from Spain, one Bell Beaker from Portugal, one Bell Beaker from Germany, and one Bell Beaker from Czechia who died in roughly the same period similarly carried K1a4a1. Four ancient K1a4a1 carriers were found in Switzerland.[404] A pre-Roman Umbri carrier of K1a4a1 who died between 768–420 B.C.E. was found at the necropolis of Plestia in East Umbria, Italy.[405] Other K1a4a1 samples come from Early Neolithic Spain, Early Neolithic Italy, Early Neolithic Serbia, Neolithic Scotland, Neolithic Morocco, the Nuragic culture of ancient Sardinia, the Únětice culture of ancient Poland, and ancient cultures from Ireland, Guernsey, and Germany.

Subclades under K1a4a1 similarly have a wide distribution. The haplogroup K1a4a1a+195 is found among the Chueta people of Spain,[406] who largely descend from Sephardic Conversos. A few of the many other examples include K1a4a1h among the Chuvash people, K1a4a1k among Berbers in Morocco, and K1a4a1b1 in the English.

Costa's team regarded their ancestral haplogroup, K1a4 (also called K1a4*), as an originally "Near Eastern" haplogroup.[407] Today, K1a4* has a wide range across Europe and West Asia, including in the United Arab Emirates, Qatar, Cyprus, Egypt, Azerbaijan, Syria, Iran, Greece, Serbia, Russia, Germany, and Portugal and in Italians, Croats, Czechs, Armenians, and other non-Jewish ethnicities. My understanding is that K1a4 is also found among Sephardic Jews from Rhodes, Greece. K1a4 was reported to be the haplogroup of four Turkish Jews, one Tunisian Jew, one Moroccan Jew, and one Libyan Jew who participated in Behar's team's research.[408] Since its subclade K1a4a has been confirmed to exist among Tunisian Jews, it is possible that a more thorough sequencing of these other North African Jews would have turned up the K1a4a-defining mutation G6260A.

Pre-modern K1a4* samples are also known. Particularly old ones are the Neolithic K1a4 from Menteşe, Turkey dating to between 6400–5600 B.C.E.,[409] the Neolithic K1a4 from Gomolava, Vojvodina, Serbia dating to between 4710–4504 B.C.E.,[410] and the Neolithic K1a4 from Isbister, Orkney, Scotland dating to between 2580–2463 B.C.E.[411] All three of those were males.

Haplogroup K1a4d is found in the Jewish community from Cochin, India but not K1a4a. Haplogroup K1a4b was found in a sample from Motza, Israel that dates from between 7300–6200 B.C.E.[412] and some Druze people (including those from Israel) have it today.

K1a9

The mtDNA haplogroup K1a9 probably originated in the northern Middle East. Costa's team acknowledged that it "might have a Near Eastern source" but still favored a more immediate source for this haplogroup in Ashkenazim in western or southwestern Europe.[413] Evidence that follows led me to disagree with their conclusion.

K1a9 is found among Ashkenazim with matrilines from Germany, Hungary, Latvia, Lithuania, Belarus, Ukraine, Moldova, Romania, Poland, Czechia, Austria, and the Netherlands. This would have been one of the original German Jewish haplogroups. In my father's list of K1a9 matches in his Full Coding Region screen inside Family Tree DNA, I confirmed that K1a9 also got assimilated into a genealogically well-documented Polish Catholic family in southeastern Poland as a result of a Jew's conversion.

K1a9 is the second most common maternal haplogroup among Ashkenazim but rare in other populations. One Kurdish person from Saqqez, the capital of Iran's Kurdistan Province, was found by Shirin Farjadian's team to be a carrier of K1a9.[414] One K1a9 carrier who tested with 23andMe has a maternal grandmother who identified as a Syrian Turkmen whose earlier matrilineal ancestors lived in Konya, Turkey until the 1800s.[415]

Doron Behar's team found Turkish Jewish, Bulgarian Jewish, Syrian Jewish, and Iraqi Jewish carriers of K1a9.[416] This builds the case for K1a9 having existed among the ancient Israelites. My father matches a person with a Jewish matriline from Antalya, a city in southwestern Turkey, at a genetic distance of 1 in his Full Coding Region screen.

An individual carrying K1a9 who claims to have a matriline from Spain (ethnicity and religion unspecified) matches some, but not all, K1a9-carrying Ashkenazim in the Full Coding Region screen. My father only matches that person in the HVR1 and HVR2 screens.

K2a

This mtDNA haplogroup is also called K2a*. Some Ashkenazim, including those with documented matrilines from Lithuania and Belarus, have been fully sequenced into K2a by Family Tree DNA rather than into its branch K2a2a1. The Ashkenazic K2a carriers who tested with Family Tree DNA match numerous non-Jewish people from western and central Europe in their Full Coding Region screen. K2a is common across modern Europe with numerous K2a carriers identified with matrilines from Germany, France, England, Ireland, and Wales and it is also found in Sweden, Switzerland, Sicily, and Spain's Basque people.

K2a was in ancient Yorkshire, England.[417]

K2a2a1

A great many Ashkenazim with matrilines from countries including the Netherlands, Germany, Austria, Czechia, Slovakia, Hungary, Romania, Poland, Ukraine, Lithuania, Latvia, and Belarus carry K2a2a's subclade K2a2a1, the fourth most common Ashkenazic mtDNA haplogroup. It is occasionally found among eastern European Christians, the K2a2a1 sample from modern Poland collected by Piotrowska-Nowak's team being an apparent example,[418] but Costa's team stated that they obtained it from Ashkenazic ancestors because its roots, K2a2a and K2a2, are only found among French and German people.[419] Due to "its nesting within a German lineage," Costa's team argued that K2a2a1 "may have been assimilated in central Europe" from an unknown convert to Judaism,[420] although they also offer the possibility that it could have been assimilated in western Europe.[421]

The real story of K2a2a1's origin is almost certainly different than what Costa's team suggested. For one thing, Behar's team studied two Mizrahi people from the Caucasus region (one a Georgian Jew, the other an Azerbaijani Jew) who also belong to varieties of K2a2a.[422] Georgian Jews and Mountain Jews from Azerbaijan do not autosomally descend from Ashkenazim. Also, Mountain Jews have no admixture from western or central Europeans at all. Another point is that other carriers of the specific branch K2a2a1 that Ashkenazim possess include people with matrilines from Italy, Portugal, Spain, and Cuba, as Leo Cooper saw inside Family Tree DNA.[423] A third point is that K2a2a1's sister subclade, K2a2a2, named by YFull's MTree, has been found in numerous people from the United Arab Emirates as well as in a Saudi Arabian.[424] I confirmed that K2a2a2 carriers share the mutation C11348T that defines membership in the K2a2a cluster and that they have their own mutation G16213A that Ashkenazim do not have.

There is one customer of Family Tree DNA with a matriline from Turkey who also carries K2a2a1 but I do not know that person's background. Some Ashkenazic Jews did move to Constantinople (Istanbul)[425] and some Turkish Jews married Ashkenazic Jews so if Costa's team had been right that would have been why K2a2a1 existed in Turkey. Ashkenazic Jews also assimilated with the Sephardic community in Bulgaria and that could be responsible for some variety of K2a2a being found in a Bulgarian Jew.[426] However, the totality of the evidence we now have led Cooper to the conclusion that Ashkenazim probably inherited K2a2a1 from the Israelites, rather than from central Europeans, and I agree with him. That would mean that K2a2a1 was probably present in Western Jewish populations in the Mediterranean region even before the original Ashkenazic-Sephardic split.

L2a1l2a

The mtDNA haplogroup L2a1l2a, defined by the mutation T14180C and the back mutation A143G!, is found in a considerable number of Ashkenazim tracing matrilines to many different places including

Germany, Austria, the Netherlands, France, Czechia, Lithuania, Latvia, Belarus, Ukraine, Hungary, Romania, and especially Poland.[427] L2a1l2a also exists among ethnic Poles[428] who descend from Ashkenazic Jews who converted to Catholicism in the mid-eighteenth century. It was probably present among the medieval Jews of the Rhineland of Germany. One of its branches, with the extra mutation C3573A, is called L2a1l2a1 and is found in Ashkenazic matrilines from Lithuania, Belarus, Poland, Ukraine, and Romania.[429] YFull's MTree estimates that L2a1l2a was formed about 8450 B.C.E. with a most recent common ancestor who lived perhaps around 550 B.C.E. or as recently as 950 C.E., and that the most recent common ancestor for all L2a1l2a1 carriers lived around 1550. The other identifiable branch, L2a1l2a2, has the mutation A12451G and its most recent common ancestor lived around 1400.

In Family Tree DNA's mtDNA Haplotree, its parent haplogroup, L2a1l2 (also known as L2a1l2*), is found in customers tracing their matrilines to places including Germany, Portugal, Morocco, and Guinea-Bissau, the last being a country in a far-western part of West Africa. Doron Behar's team collected a sample from a Mandinga person from Guinea-Bissau[430] that is assigned to L2a1l2* by YFull's MTree. L2a1l2* was also found in three Gambian people who participated in the 1000 Genomes Project[431] and in a Gambian from the Fula ethnicity who participates in YFull's MTree.[432] The Gambia is another country in a far-western part of West Africa. Also notable are the two people from Burkina Faso who belong to L2a1l2*; at least one of them is a member of the Pana ethnicity.[433] Burkina Faso is a landlocked country in West Africa. Here we see links to the ultimately Sub-Saharan African origins of the entire L2a1 family of haplogroups.

The deeper ancestral haplogroup L2a1l, in turn, is found in a handful of Family Tree DNA's customers with matrilines from France, Algeria, and Haiti, with the last of these lineages obviously stemming previously from Sub-Saharan Africa before the 1600s. L2a1l has also been found by scientists in modern people from Sierra Leone and The Gambia. L2a1l was also the haplogroup assignment of a person who lived in what is now Córdoba, Spain circa 1580 B.C.E.[434]

Related subclades of L2a1l are also found in Sub-Saharan Africa. Haplogroups L2a1l1 and L2a1l1a are both found in the Nuna and Mossi peoples of Burkina Faso and Nuna people also carry L2a1l1b. L2a1l1a1 is found in Bissa people and L2a1l1a2 in Marka people; those are ethnic groups in Burkina Faso and neighboring countries. L2a1l3 is found in Burkina Faso, Nigeria, and Algeria.

Given what we know of Jewish migrational history and autosomal DNA, Ashkenazim would have inherited L2a1l2a directly from a North African Berber female rather than directly from a West African female. In reliable tests that measure ancestry within the past several thousand years, Ashkenazim do not score any West African autosomal DNA but do score minor North African. An Iberian female could not be the haplogroup's source for medieval German Jews.

Some Ethiopian Jews belong to the distantly related haplogroup L2a1b2[435] but they are a population that entirely descends from Ethiopian converts to Judaism.

M1a1b1c

The mtDNA haplogroup M1a1b1c exists among Ashkenazim with matrilines from Germany and Poland. It belongs to a family of lineages that spans from North Africa to southern Europe and the Middle East. Its parent haplogroup, M1a1b1, has been found among people in Tunisia and Spain. Its sister haplogroup, M1a1b1a, is prevalent in Italy, including on the Italian island of Sardinia, and that was the reason that Costa's team wrote that M1a1b "is characteristic of the north Mediterranean and was most likely assimilated there" by Ashkenazim.[436] Another sister haplogroup, M1a1b1b, has been found in Yemen, the United Arab Emirates, Georgia, and Ossetia, and a daughter branch of that known as M1a1b1b1 is found in Yemen and Saudi Arabia.

Hungary's Püspökladány-Eperjesvölgy cemetery, containing burials from the tenth and eleventh centuries, included a female whose haplogroup was sequenced as M1a1b1.[437]

A variety of M1a1 that is possibly of Judean origin is found among Moroccan Jews. Some type of M1a1 is also found in Yemenite Jews, potentially equal to one of the Yemenite Arab subclades of M1a1 but I do not know at this time. M1a1 is found among both Ethiopian Jews and Ethiopian non-Jews; it is known from autosomal DNA that Ethiopian Jews are not descended from the Israelites.

M33c

M33c is an East Asian lineage that connects Eastern Ashkenazim to the Chinese. It is found in Ashkenazic populations from east-central and eastern Europe, including those of Lithuania, Latvia, Belarus, Ukraine, Poland, Romania, and Hungary. Small proportions of Austrian Jews and French Jews also had it. Indications are that M33c would not have been found among the medieval German Jews.

The genetically most similar sequence to the Ashkenazic sequences was found in a Han Chinese person from the province of Sichuan in southwestern China.[438] The branch of M33c to which all of the Ashkenazim and this Sichuanese individual belong was tentatively called M33c2 by Tian's research team and it is primarily defined by the presence of the mutation C4182T. However, M33c2 is not a recognized subclade in GenBank nor in Mannis van Oven's phylogenic tree. YFull's MTree uses a different nomenclature so that MTree's M33c2 is something different than the M33c2 branch identified by Tian's team even though it is also Chinese. In MTree, the Ashkenazic branch is called M33c3 instead.

More distant matches to this subclade are found among six ethnic groups in China (Han, Zhuang, Yao, Miao, Kam-Tai, and Tibetan) as well as among Han in Taiwan and occasionally among a few ethnic groups in Thailand and Vietnam. They do not belong specifically to M33c2 although they are within the parent haplogroup M33c's cluster, with some of them being classified as members of tentatively named subclades M33c1, M33c3, M33c4, and M33c5 by Tian's team. Most of M33c's genetic diversity is found in China, which strongly suggests China as the point of origin for the first M33c-carrying woman. The Zhuang (Cuengh) people live in the Guangxi Zhuang Autonomous Region in southern China. Yao people live in southern and southwestern China in the provinces of Guangxi, Guangdong, Guizhou, and Yunnan as well as in Myanmar and northern areas of Vietnam, Laos, and Thailand. They are called Dao in Vietnam. Miao people are concentrated in China's Guizhou province but other Miao live in the provinces of Yunnan, Sichuan, Guangxi, Hunan, Hubei, and Hainan so they are found in both southern and southwestern China, and the Hmong branch of the Miao migrated into many countries in southeast Asia (Myanmar, Laos, Thailand, and northern Vietnam). Kam-Tai is a linguistic term that encompasses languages spoken by Kam-Sui and Tai (Dai) ethnicities in southern China and southeast Asia, including Thailand. Tibetans live not only in Tibet but in western Sichuan, directly east of Tibet.

Tian's team determined through mitochondrial analysis that the woman carrying M33c2 joined the Eastern Ashkenazic community during the Middle Ages and no later than the year 1375. From what we currently know, M33c2 originated in southern China, probably southwestern China, and the first Jewish M33c2 carrier probably lived in the 1200s or 1300s and migrated along the Silk Road to eastern Europe, becoming one of Europe's earliest permanent residents of Chinese origin.

There is no evidence to connect M33c to the centuries-old Jewish community of Kaifeng, China, originally of Persian Jewish origin but later largely Han.

Indications are that M33c's matrilineal ancestors originally migrated into China from India. M33c's parent haplogroup M33-T16362C (called M33-a* by YFull's MTree) was found in a person from northern India by a medical research team led by D. Ramanan.[439] M33c and its sister haplogroup M33b share the T16362C mutation with their parent. M33b is present in Nepal[440] and Myanmar[441] while its subclade M33b1 is found in Tibet, China, and Myanmar and its subclade M33b2 is in Nepal and India.[442] YFull's MTree identified a new subclade called M33b2a that is in one of their Indian customers.

M33-T16362C is a daughter subclade of M33, which has other varieties that are found in South Asia. M33a is found among the Bhil tribe in India's northwestern state of Gujarat and among the Khasi and Garo peoples of northeastern India and Bangladesh. Its branch M33a1a is found in Nepal and among Lepcha tribes in Sikkim in northeastern India. M33a1b30 is present among the Dongri Bhill tribe from Madhya Pradesh in central India and Rajasthan in northwestern India. M33a2 is found in tribes in Maharashtra in western India as well as some members of the Rajbanshi tribe in West Bengal and among Brahmins in Uttar Pradesh in northern India. M33a2'3, M33a3, and M33d are also found in India.

N1b1a2

N1b1a2 is a very rare mtDNA haplogroup in the Ashkenazic population. I confirmed that N1b1a2 is carried by a member of an Orthodox Jewish family who is genetically fully Ashkenazic in his recent ancestry (per

Eurogenes Jtest's oracle) and whose documented matriline traces back to a Jewish woman who lived in a town in East Prussia that is today in northeastern Poland. EMPOP's database includes a N1b1a2 carrier who declared an Ashkenazic matriline from Hungary.

N1b1a2 also has associations with Sephardic descendants and with modern and ancient peoples of the Levant region. The Crypto-Jews in the Bragança district in northeastern Portugal descend from Sephardic Jews. In EMPOP's database, Inês Nogueiro's research team found N1b1a2 in two members of Bragança's Crypto-Jewish community: one from Zamora and one from Miranda do Douro, "both places with a well-documented history of Jewish presence."[443]

N1b1a2 is found in Lebanon, Syria, Yemen, Saudi Arabia, Qatar, Bahrain, and the United Arab Emirates, in Armenians from Armenia and Turkey, and among the Georgian and Assyrian peoples. N1b1a2 has been found in three ancient samples from the Levant, including a female found at Tel Megiddo in Israel who died between 1900–1700 B.C.E., a male found in Yehud, Israel who died between 2500–2000 B.C.E., and a male found in Jordan who died between 1342–1164 B.C.E.[444] An ancient N1b1a2 carrier was found at the Arslantepe archaeological site in Turkey's Malatya Province.[445] N1b1a2 was also found in three individuals from Christian-era Nubia in Sudan. The subclade N1b1a2d is found in the United Arab Emirates and in Armenians from Turkey.

However, N1b1a2 is also found in modern non-Jewish matrilines from Europe including but not limited to Italians, Irish, and Germans. Piotrowska-Nowak's team found it in a modern person from Poland who is probably not Ashkenazic[446] and also found the subclade N1b1a2a in Poland.[447] Other scientific teams found the subclade N1b1a2b in Poland and Russia. N1b1a2 was also in medieval Hungary. A female carrier of N1b1a2 was buried at the Vagnari necropolis in southeastern Italy sometime between the first-fourth centuries when it was part of the Roman Empire.[448]

Ashkenazim probably inherited N1b1a2 from the ancient Israelites, but it is not impossible that they got it from a European ancestor instead.

N1b1b1

N1b1b1 is the third most common mtDNA haplogroup among Ashkenazim after K1a1b1a and K1a9. It was previously called N1b2 and is still called N1b2 by 23andMe. (The current N1b2 is something different and by coincidence it is also found among Ashkenazim, as I will discuss in the next entry.) N1b1b1 is defined by the presence of the mutation T14581C. N1b1b1's carriers include Ashkenazim from Germany (including Bavaria and Baden-Württemberg), Austria, the Netherlands, Belgium, France, Switzerland, Hungary, Czechia, Lithuania, Latvia, Belarus, Romania, Moldova, Slovakia, and especially Poland and Ukraine.[449] There is every reason to believe that medieval German Jews also carried it.

Ashkenazic N1b1b1 carriers who tested with Family Tree DNA match many people of definite and potential Sephardic matrilineal descent. In the Full Coding Region screen, some of them have a genetic distance of 1 from a N1b1b1 carrier with a Sephardic Jewish matriline from Bulgaria and from a N1b1b1 carrier with a Sephardic Jewish matriline from Rhodes, Greece.[450] A second person with a Sephardic Jewish matriline from Greece (and Turkey) appears as a match only in the HVR1 screen because he only ordered a partial mtDNA sequence (showing with the prediction "N1b"). Three additional people with matrilines from Bulgaria show as HVR2-level matches. Other HVR2-level matches include a person with a Moroccan Jewish matriline listed as Sephardic, a person with an Egyptian Jewish matriline, and three people with Turkish Jewish matrilines. Some of the Ashkenazim have a genetic distance of 2 from a person with a matriline from Italy and there are two more matches with matrilines from Italy at the HVR2 level plus one more at the HVR1 level. There is an HVR2-level match to a person with a matrilineal ancestor from Jamaica who had a non-Jewish surname.

Unexpected exact matches to some of the Ashkenazim include two non-Jews with early American roots: one had a matrilineal ancestor born in 1750s Kentucky and the other had a matrilineal ancestor born in Kentucky in 1817, and both of those ancestors had Northwestern European Christian first and last names. A person with a genetic distance of 1 from some of the Ashkenazim had a matrilineal ancestor with English first and last names who was born in 1777 in a place not specified. A person with a matriline of unspecified ethnicity from the country of Georgia matches some of the Ashkenazim at a genetic distance of 2.

N1b1b1 was the haplogroup assignment for two sixth-century Germanic Langobard (Lombard) samples that were found at Szólád,

Somogy County, Hungary.[451] Neither of these Langobard N1b1b1 carriers showed any autosomal admixture from the Middle East. N1b1b1 was also the haplogroup of a sixth-century Langobard from the Mušov-Roviny burial site in Czechia.[452]

Its parent haplogroup, N1b1b, is found among modern Sardinians, Belgians, and the Irish and people from Austria, Germany, and Poland. N1b1b was also found in two ancient samples from Switzerland.[453] A female N1b1b carrier who died circa 1450–1250 B.C.E. was found at Tel Hazor in Israel's Upper Galilee region.[454]

Some other branches of N1b have a strong presence in the Middle East. As stated above, N1b1a2 is found in many modern populations in West Asia and Southwest Asia from Anatolia and the southern Caucasus in the north to Yemen in the south. N1b1a3 is found in countries like Saudi Arabia and Lebanon. N1b1a7 is found in Yemen, Turkey, Armenia, Spain, and among the Assyrians. Therefore, judging by these other branches, the root haplotype N1b is evidently of Middle Eastern origin, but this turns out not to be the immediate case for the Ashkenazic branch, which was described as a "putative west Mediterranean" lineage by Costa's team.[455] The existence of numerous Sephardic matches (including close ones) from so many countries to the Ashkenazim appears to rule out a central European originator of this haplogroup for the Ashkenazim. Southern Europe, perhaps Italy, was probably their source instead.

N1b2

I confirmed that N1b2 is the current mtDNA haplogroup assignment for at least three full Ashkenazim who tested their complete sequences with Family Tree DNA. At least one of them has a documented Galitzianer Jewish matriline from Ukraine.[456] Family Tree DNA's mtDNA Haplotree also lists a N1b2 carrier with a matriline from Germany. This haplogroup, which includes the mutations C5507A, A12026G, T15784C, and A16258C, appears to originate from the Middle East. It is also found among the non-Jewish inhabitants of Yemen[457] and Somalia. YFull's MTree was the first to name a branch of N1b2 that they call N1b2a. MTree lists N1b2* (equivalent to N1b2) as being present in Somalia and N1b2a in both Somalia and Yemen. The published Yemenite sample belongs to N1b2a, which is defined by the mutation G7013A.

N9a3

The East Asian mtDNA haplogroup N9a3 has a low frequency among Eastern Ashkenazim. It is occasionally encountered among Ashkenazim with matrilines from Lithuania, Belarus, Ukraine, Hungary, and northeastern Poland.[458] Indications are that N9a3 would not have been found among the early medieval German Jews. N9a3 apparently entered the Turkish Jewish population as the result of an Ashkenazic migration to Turkey.

An Ashkenazic N9a3-bearer who tested with Family Tree DNA up to the Full Coding Region level matches three notable non-Jews with a genetic distance of 3: one Ingush, one Chechen, and one Chinese.[459] The first two are from the Caucasus region, while the Chinese match could suggest a migration path for Ashkenazim across the Silk Road directly from China, as with M33c. At the HVR1 level, the Ashkenazim match several more Chechens plus another Chinese individual (from southeastern China).[460]

Haplogroup N9a3 is common among Han Chinese people from Shanghai and Qingdao in eastern China and from Taiwan.[461] As of January 2022, YFull's MTree classification system includes Taiwanese people in the subclades N9a3a1a and N9a3a1a1. Other ethnic groups among which N9a3 is present include Koreans from South Korea,[462] Japanese, Buryats from South Siberia,[463] Uyghurs, and Uzbeks. It is also found in Southeast Asia (Thailand and/or Laos) and among Pamir highlanders of Central Asia.

A medieval carrier of N9a3 was found among the Hanging Coffin samples from Yunnan province in southwestern China.[464] YFull's MTree places this sample into the subclade N9a3a1*.

A branch of N9a3 called N9a3a is found very infrequently in modern people from central and eastern Europe. An ethnic Russian from Belgorod in western Russia and an ethnic Czech from West Bohemia are both assigned N9a3a in Derenko's team's study and are listed that way in GenBank.[465] YFull's MTree classifies these Russian and Czech samples as carriers of a subclade they call N9a3a3a, defined by the presence of the mutation C12636T. N9a3a is also in Chechnya, South Korea, and Japan.

YFull's MTree says its newly named subclade N9a3a1b* (formerly N9a3a2*) is defined by the mutation C7810T and places two ethnic

Bashkir people from Bashkortostan, Russia into it.[466] The Ashkenazic carriers of N9a3 who tested their mtDNA with Family Tree DNA show the same mutation (C7810T) in their Coding Region. According to Leo Cooper, based on Behar's team's earlier research, the Ashkenazim and Sephardim belong to a branch of that called N9a3a1b1 (formerly N9a3a2a) by MTree, which is defined by the back mutation T146C!.[467] This is because the Ashkenazim had the mutation C146T reversed. At least some Southeast Chinese and Chechen carriers of N9a3 at Family Tree DNA have kept C146T, as has the Ingush carrier there. At this time, I do not have complete lists of mutations for the Chechen, Ingush, and Chinese carriers who are close matches to the Ashkenazim, but YFull's MTree has identified two Ingushes and three Chechens at the root level of N9a3a* which is defined by the mutation T12354C.

Pre-modern examples of N9a in Europe were the subject of controversy. Once thought to be from the 6th millennium B.C.E. during the Neolithic era, a fresh analysis by a different set of scientists established that in fact they were not nearly as old. Bánffy's team explained that these N9a carriers were actually two Sarmatians who lived in the classical era and one medieval Hungarian who lived soon after the Magyar conquest of the Carpathian Basin in 896.[468] Sarmatians and Magyars both migrated into Hungary from lands to the east.

An ancient female member of the Xiongnu people of the eastern Eurasian steppelands was found to belong to the parent haplogroup N9a*.[469]

Because N9a3 has not been found in non-Jewish ethnic groups of Europe who are known to have intermarried with Ashkenazim, such as Poles and Germans, I am inclined to believe that the first Jewish N9a3 was a Chinese, North Caucasian, or Khazarian convert to Judaism. Current phylogenetic analysis hints at a Khazarian connection.

R0a2m

The Middle Eastern mtDNA haplogroup R0a2m is found in many Ashkenazic matrilines, including some from Germany, Czechia's Moravia region, Poland, Ukraine, Belarus, Lithuania, Latvia, and the Netherlands. Within Family Tree DNA's Full Coding Region screen, many (but not all) of these Ashkenazim are exact mtDNA matches to an Ecuadorian and a Moroccan who also have R0a2m as a terminal haplogroup assignment. In addition, many of these Ashkenazim have a genetic distance of 1 from a Mexican and from a Chueta from Mallorca, Spain and both of those people are also assigned to R0a2m.[470] Many of the Chuetas' ancestors were Sephardic Conversos. The Ecuadorian match has documentation that traces her direct maternal line to the Catholic woman Juana Rodriguez Carreño, who was born in 1512 in Badajoz in western Spain's Extremadura region.[471] Sephardic Jewish women from the 1300s-1400s would be the ultimate source of the R0a2m lineages of these Ecuadorian, Mexican, Chueta, and Moroccan people.

It is less clear whether any or all of the Ashkenazim possessing R0a2m had distant Sephardic ancestors in their direct maternal lines. The closeness of the aforementioned matches had suggested that possibility to me. However, R0a2m is widespread among Ashkenazim, and one of the Ashkenazic R0a2m carriers with roots in Moravia is closer to some of the Hispanic and North African matches than to some of the Ashkenazic matches, connecting to certain Ashkenazim at genetic distances of 2 and 3. Perhaps R0a2m entered the Ashkenazic population more than once, such as first through a medieval Italian Jewish woman who arrived in Germany and second through an early modern Sephardic immigrant to, perhaps, Czechia.

Their other mtDNA matches with provable Jewish connections, who do not show in the Full Coding Region screen, include Sephardic Jews from Morocco, Tunisia, Libya, Bulgaria, and Syria and two Chuetas from Mallorca who are descendants of Sephardic Conversos. The Chuetas, like Ashkenazim, have also been assigned to the specific haplogroup R0a2m, and it is their most frequent maternal haplogroup.[472] The Libyan, Tunisian, and Syrian matches who explicitly declare matrilineal Sephardic ancestry match at least some Ashkenazim on the HVR2 screen and their haplogroup is listed simply as R0a, apparently because they did not have their complete sequences tested. The Moroccan Jewish and Bulgarian Jewish matches only show as matches to some Ashkenazim on the HVR1 screen and their haplogroup is likewise listed simply as R0a.

My understanding is that the pre-2014 name for R0a2m was R0a1a. Costa's team described R0a1a as the most common R0a variety in Ashkenazim and Behar's team noted that R0a1a is one of the four major mtDNA haplogroups in Tunisian Jews.[473] This implies that if the two "R0a" Tunisian Jews in Family Tree DNA were to test their complete mtDNA sequences, their terminal haplogroup would display as R0a2m.

Other distant "R0a" matches to the Ashkenazim at only the HVR2 level include one Bedouin from Israel, one Palestinian Arab, one person with a matriline from Iran, and two people with matrilines from Italy.

The indications are that R0a2m would have been a lineage among the ancient Israelites. Before then, according to Francesca Gandini's team, the ancestors of the Jewish R0a2m women would have lived in Arabia, consistent with the origins of its ancestral haplogroups R0a2 and R0a.[474] R0a2 is found in many modern Middle Eastern populations, including Yemenis, Saudi Arabians, Emiratis, Qataris, Kuwaitis, Assyrians, Druze, Palestinian Arabs, Bedouins, and Iranians but also in some areas of East Africa, southern Europe, and India. Many branches of R0a2 are also common in the Middle East. R0a2c is found in Yemen, Saudi Arabia, Kuwait, Qatar, the United Arab Emirates, and Lebanon and among Bedouins. R0a2h is found in Yemen, Oman, and the United Arab Emirates. R0a2j is found in Yemen and the United Arab Emirates and among Armenians. Other branches found in Yemen include R0a2b1b1, R0a2f1a, R0a2g, R0a2k1, and R0a2l.

However, Leo Cooper argues that R0a2 and R0a did not necessarily originate in Arabia. Cooper notes that the modern spread and density of these haplogroups may be misleading when searching for their origins. Archaeologists found several quite ancient R0a2 and R0a samples in the Levant region. A female R0a2 carrier who died between 6800–6700 B.C.E. was found at the Pre-Pottery Neolithic archaeological site ʿAin Ghazal in northern Jordan.[475] A male R0a carrier who died between 7722–7541 B.C.E. was also found at ʿAin Ghazal.[476] A female R0a carrier who died between 4500–3900 B.C.E., during the Chalcolithic era (Copper Age), was buried in Peqiʿin Cave in Upper Galilee, Israel.[477] Judging by the ancient Levantine R0a carriers, Cooper suggests that R0a arrived in Arabia from the north.

R0a4

YFull's MTree estimates that R0a4 is about 15,000 years old. R0a4 has been confirmed to be a mtDNA haplogroup among Ashkenazim,[478] including those with matrilines from the Netherlands, Germany, France, Hungary,

Romania, Poland, Lithuania, Belarus, Ukraine, and Russia proper. It is rare in non-Ashkenazic populations. R0a4 carriers are known from Spain and Iraq.[479] At least some Ashkenazic R0a4 carriers who tested with Family Tree DNA can see three individuals with Spanish-language names in their Full Coding Region screen, two as exact matches and one with a genetic distance of 1.[480] More distantly, in the HVR2 screen, Ashkenazim have two Bulgarian Jewish matches.

T1a

T1a, also called T1a*, is an uncommon mtDNA haplogroup among Ashkenazim today. It is the confirmed complete sequence assignment for multiple people whose documented matrilines trace back to Ashkenazic women from Germany. I received information that one of these T1a Ashkenazim has a direct maternal line from Baden region near the Upper Rhine in southwestern Germany going back at least to the start of the 1800s.

The German Jewish T1a carriers distantly match two other T1a carriers with Arabic names in the HVR1 screen. As of the time of this writing, Family Tree DNA's mtDNA Haplotree includes one T1a carrier with a matriline from Iraq and two with matrilines from Sudan plus three with matrilines from Turkey, one from Armenia, and one from Israel who may or may not be Jewish. It also lists many T1a carriers with matrilines from European countries including Portugal, Spain, Germany, Poland, Russia, Ukraine, Slovakia, Hungary, Croatia, Italy, France, Belgium, England, and Ireland, among others, and nearly all of these people must be non-Jews. A T1a carrier who tested with Family Tree DNA and then uploaded the mitochondrial data to GenBank has a matriline listed as ethnic Bulgarian. Another T1a carrier who shows as a match in the HVR1 screen has Greek first and last names. YFull's MTree lists T1a* carriers in countries like Denmark and Tunisia along with a carrier whose ethnicity is identified as French by the Human Genome Diversity Project.[481]

A very ancient female carrier of haplogroup T1a was found at the Pre-Pottery Neolithic archaeological site 'Ain Ghazal in Jordan and she died between 7446–7058 B.C.E.[482]

T1a1

I received confirmation, and later confirmed myself, that some full Ashkenazim who tested their complete mitochondrial sequences with Family Tree DNA were assigned to the haplogroup T1a1 and that it therefore represents an ancestor who was distinct from the ancestors who belonged to T1a, T1a1b, T1a1j, and T1a1k1. Some of the carriers of T1a1 have documented Ashkenazic matrilines from Ukraine and Poland.

This haplogroup is especially widespread in European ethnic groups. Published studies found modern carriers of T1a1 with non-Jewish matrilines from such countries as Finland, Russia, Poland, Bulgaria, Italy (including Apulia and Sardinia), Ireland, and England and among the Basques but also from Asia including in the Altai region, the Pamir region, the Uyghurs, Tajiks, Pathans, Kashmiris, and Zoroastrian Parsis from India. Participants in Family Tree DNA who were sequenced as T1a1 include ethnic Finns, Ukrainians, Poles, and Armenians and people from Switzerland, Wales, Estonia, Greece, Qatar, Bahrain, Syria, Lebanon, Kazakhstan, and Uzbekistan, among many other ethnicities and countries.

The Ashkenazic carriers of T1a1 match northern and central Europeans particularly closely and numerously, suggesting that a European convert to Judaism, perhaps a German, introduced T1a1 into the Ashkenazic community. In Family Tree DNA's Full Coding Region screen, one of the Ashkenazic testers has a genetic distance of 1 from Norwegian, Swedish, German, English, and Irish matrilines plus a few from Ukraine, one from Serbia, one from Scotland, one from northern France, and one from Portugal's Madeira region, a genetic distance of 2 from Finnish, Irish, English, Scottish, Slovakian, and Croatian matrilines, and a genetic distance of 3 from Polish, Norwegian, German, and Cornwall matrilines, among others. All of them were assigned the haplogroup T1a1. The Ashkenazic participants do not all match each other closely; some are exact matches while others have genetic distances of 1, 2, and 3 from some of the others, reflecting the existence of random mutations in the Ashkenazic subset of T1a1.

T1a1 is known to have been present in medieval times in Denmark,[483] England,[484] Hungary,[485] and Russia's Ural region.[486] A Germanic Langobard carrier of T1a1 was also found.[487]

There are abundant ancient T1a1 samples as well, including from the Neolithic, Copper, and Bronze Ages. A female T1a1 carrier dating back to between 2564–2475 B.C.E. was found in Karsdorf, Germany.[488] Two Bell Beakers were found in Germany's Bavaria region dating to between 2500–2000 B.C.E.[489] A female T1a1 who died circa 1881–1701 B.C.E. was found underneath Prague, Czechia.[490] A male T1a1 who died circa 1208–978 B.C.E. was found in Bedfordshire, England.[491] A male member of the Corded Ware culture with this haplogroup was found in the Malopolska Upland in southern Poland.[492] A female T1a1 from the Yamnaya culture was found on the Pontic steppe of Ukraine and her remains date to between 3300–2700 B.C.E.[493] T1a1 was also in ancient Mongolia.

T1a1b

Inside Family Tree DNA, I was able to confirm that T1a1b is a mtDNA haplogroup that full Ashkenazim can carry, albeit very rarely. YFull's MTree estimates it formed about 5,100 years ago. T1a1b also exists in ethnic Poles and Czechs along with modern people from other European countries like Denmark, Sweden, Finland, and Bulgaria. In West Asia, a particular variety of it that MTree named T1a1b2b is found among the Lebanese and Anatolian Turks. Further east, T1a1b has been detected in Buryat people from Siberia and Pamir highlanders in Central Asia. At this time, I do not know whether the Ashkenazic branch is closer to a European branch or to a West Asian branch. According to Wim Penninx, the Ashkenazic branch carries the mutation G16000A; as of January 2022, this mutation is not listed for any of the non-Ashkenazic T1a1b carriers in the mtDNA databases maintained by Ian Logan and YFull.

Multiple old carriers of T1a1b have been unearthed in Europe. A T1a1b carrier was buried in Denmark's rural Refshale site during the middle of the Black Death period.[494] A male sample from Karos-Eperjesszög cemetery number three from ninth or tenth century Hungary also belonged to T1a1b.[495] Its mitochondrial sequence is identical to that of an Avar male who lived in the land that became Hungary in the seventh century, that is, prior to the Magyars' conquest of the land.[496] An even older T1a1b carrier, a male from circa 800–545 B.C.E. during the Eastern Baltic Bronze Age, was found in Ķivutkalns, Latvia.[497]

T1a1j

Aside from Ashkenazic Jews, such as those with matrilines from Austria and Ukraine, the mtDNA haplogroup T1a1j is also found among Sephardic Jews from the Greek island of Rhodes[498] and from Milas in southwestern Turkey. At least some of the Ashkenazim who tested with Family Tree DNA can see the Turkish Jew from Milas matching them in the Full Coding Region screen with a genetic distance of 2.[499] Several more T1a1j customers of Family Tree DNA likewise have Sephardic matrilines.

In addition, T1a1j is carried by a non-Jew from Sweden who took the mtDNA test from Family Tree DNA and matches at least some of the Ashkenazim with a genetic distance of 3. We know that T1a1j was already present in medieval Sweden because a male dug up from the Nunnan cemetery in Sigtuna, Sweden was sequenced with this haplogroup.[500]

Ashkenazim, the Sephardim from Rhodes, and the Swede do not carry the mutation C5839T that YFull's MTree uses to define the novel subclade T1a1j1.

T1a1k1

T1a1k1 is another confirmed variety of T1a1 among Ashkenazim[501] and it was found in Jews in places including Lithuania, Romania, Bessarabia, Poland, Ukraine, and Belarus. It appears that Ashkenazim acquired this haplogroup from a northwestern European convert to Judaism. At least some of the Ashkenazic T1a1k1 carriers who tested with Family Tree DNA have two ethnic Dutch people appearing in their Full Coding Region screen at a genetic distance of 1.[502] In addition, they match two American non-Jewish individuals with matrilines from eighteenth-century Pennsylvania and Virginia; one of them has a genetic distance of 1 while the other has a genetic distance of 2. A T1a1k1 carrier with a matriline of unlisted ethnicity from Baden-Württemberg, Germany tested with YFull.

T1a1k1 formed about 3,300 years ago according to YFull's MTree.

Its sister subclade T1a1k2 as well as their parent haplogroup T1a1k are both found in Denmark. A late-medieval T1a1k sample was unearthed in England that dates to after the Black Death period.[503]

T1b

The mtDNA haplogroup T1b, also called T1b*, is found in a very low percentage of Ashkenazim. This has been confirmed through full sequence testing at Family Tree DNA. Most of their maternal lines come from Ashkenazic women from Ukraine but a few others trace back to Ashkenazic women from Poland and Latvia.

Outside of Jewish communities, T1b* has been fully sequenced in modern people from Saudi Arabia, Iran, Egypt, Tunisia, and Austria and in Brahmins from West Bengal, India. In addition, according to the traditional scientific nomenclature (van Oven's phylogenic tree), T1b is found among such ethnic groups as Palestinian Arabs, Uyghurs, Armenians from Armenia and Turkey, and Italians from Venice, the region of Umbria, and the province of Potenza. However, YFull's MTree places the Palestinian and Armenian carriers into a new subclade of T1b that they call T1b5.

An infant who was buried at the Arslantepe archaeological site in Turkey's Malatya Province between 3941–3708 B.C.E. was placed into haplogroup T1b.[504]

T1b in Ashkenazim could perhaps represent an inheritance from an ancient Israelite woman, if not an Italian woman.

T1b3

Costa's team indicated that T1b3 is "likely of . . . Near Eastern origin"[505] and this appears to be correct because T1b3 and some of its sister subclades as well as its ancestral form T1b are found among Middle Eastern peoples. YFull's MTree estimates T1b3 is about 9,300 years old.

Among Ashkenazim, some T1b3 matrilines trace back to Ukraine, Belarus, and Lithuania. At least some of the Ashkenazic T1b3 carriers who tested with Family Tree DNA are exact matches in the Full Coding Region screen to a Sephardic Jew from Turkey whose matrilineal ancestry comes from Turkey and Bulgaria.

Outside of Jewish communities, the root level of T1b3 is found among non-Jewish inhabitants of Turkey, Greece, and Iran, and daughter branches of it, defined by extra mutations, are found in Turkey, at least among that country's Kurds and Armenians, and in the North Caucasus region of southern Russia.[506]

The wider family of T1b subclades continues the strong association with West Asia because T1b3's sister haplogroup T1b4 is found in Syria, among Armenians, and in the Umbria region of Italy, and T1b3's sister haplogroup T1b1 is found in Jordan and Georgia but also in Finland. A subclade first identified by YFull's MTree is T1b7, which is found in Turkey.

T2a1

Haplogroup T2a1, also called T2a1*, has been fully sequenced in a low proportion of full Ashkenazim who tested with Family Tree DNA, and at least some of their matrilines trace back to Ukraine and Lithuania. Family Tree DNA's mtDNA Haplotree also lists many T2a1 carriers from other European countries, including Poland, Belarus, Slovakia, Czechia, Austria, Germany, Sweden, Finland, Bulgaria, Croatia, Slovenia, and Italy, and some of those people are not Ashkenazim. The matches to the Ashkenazic testers in their Full Coding Region screen include people from east-central and eastern Europe with Ukrainian and Hungarian names.

Although T2a1 probably arrived in the Ashkenazic population from a European convert to Judaism, this haplogroup is also found in West Asia. The Haplotree lists a T2a1 participant from Georgia and scientific studies fully sequenced it in an Iraqi and an Armenian.

There are abundant daughter subclades of T2a1 and they are found in numerous peoples of Europe, North Africa, West Asia, and Central Asia and among China's Uyghurs.

T2a1b

Some Ashkenazim with matrilines from Ukraine are members of mtDNA haplogroup T2a1b, which is about 8,000 years old according to YFull's MTree. Costa's team posited that T2a1b is "likely of European . . . origin" for Ashkenazim.[507] Several people from Greece have also been confirmed to belong to T2a1b. One of them, from the Greek island of Chios, matches some Ashkenazim inside Family Tree DNA's Full Coding Region screen with a genetic distance of 2.[508] T2a1b has also been found in a Pamir highlander in Central Asia and individuals from Italy's Umbria region, Poland, and Finland.

Descendant branches of T2a1b are found in far-flung areas of modern Europe, West Asia, and North Africa. T2a1b1 is found in Poland and Sardinia. T2a1b1a is found in Portugal, Italy, Poland, and among the Irish but also among Assyrians, Samaritans, and Palestinian Arabs and residents of Jordan and Turkey. T2a1b1a1 is found in countries like Denmark, Sweden, Poland, Romania, Azerbaijan, and Turkey. T2a1b1a1b is likewise found in Denmark and Poland but also in Tunisia. T2a1b2b is present among Kurds from Georgia, the Adygei (western Circassians) of the North Caucasus, and in the United Arab Emirates.

Ancient examples of this haplogroup family are also known. A female from the Yamnaya culture who was buried in Golyamata Mogila, Popovo, southeastern Bulgaria circa 2986–2486 B.C.E. and was identified as simply a carrier of undifferentiated haplogroup T by Wilde's scientific team[509] turned out to belong within the subclade T2a1b and later more specifically to the terminal subclade T2a1b1a, as was determined by outside researchers. A female carrier of haplogroup T2a1b1 from the Corded Ware culture who died sometime between 2454–2291 B.C.E. was excavated from Esperstedt, Germany.[510] A T2a1b1 carrier was buried at the Roman-era Vagnari necropolis in southeastern Italy between the first-fourth centuries.[511]

Ashkenazic T2a1b carriers possess the extra mutation A1530G that MTree uses to define the novel subclade T2a1b4b that is estimated to originate about 2,100 years ago. A scientific sample[512] and a YFull customer, both from Poland and lacking an indication of Ashkenazic origin, are in this subclade, so a Polish origin for Ashkenazic T2a1b carriers is possible. Its parent level, T2a1b4, is found in Finland and Greece.

T2b3-C151T

This is an extremely rare haplogroup among Ashkenazim. It is a variety of T2b3 with an extra mutation called C151T. Inside Family Tree DNA, I was able to confirm that T2b3-C151T (also called T2b3+151) is the complete mitochondrial sequence held by one person who is fully Ashkenazic in recent generations both genetically and genealogically and has a matriline from Moldova.

In non-Jewish groups, T2b3-C151T is particularly common among French people, including those from the Normandy region in northern France. The haplogroup also has significant presences in England and Germany, according to Family Tree DNA's mtDNA Haplotree. As we learn from the Haplotree and published scientific studies, there are also carriers with matrilines from other European countries, including Portugal, Spain, Ireland, Scotland, the Netherlands, Denmark, Switzerland, Serbia, Greece, and Italy (including among the Sardinians), but also from Jordan, Turkey, and Morocco and in Azeri people.

A carrier of T2b3-C151T was buried at the Kelif el Boroud archaeological site in Morocco, which dates to between 3780–3650 B.C.E.[513] Two Early Neolithic T2b3-C151T carriers were found in Spain's El Toro cave, where the human remains date to between 5280–4780 B.C.E.[514] Another ancient T2b3-C151T person was found in an Iron Age necropolis at the archaeological site of Polizzello in Sicily.[515]

In addition, a person who died between the fourth-first centuries B.C.E. and was buried in a tomb at the "BEY 197" archaeological site in Beirut, Lebanon was a carrier of T2b3-C151T.[516] However, this was not a Phoenician person. Matisoo-Smith's team wrote that the existence of this haplogroup in ancient Lebanon was

> unusual, as it does not appear to be an indigenous lineage. The tomb in which the burial was found was not marked and did not have any of the artefacts found in site 198, suggesting different burial customs. This is further evidence suggesting that this individual belonged to a different socioeconomic class or ethnic group, perhaps a slave brought to Beirut from one of the Phoenician settlements.[517]

They believe that additional supporting evidence for their hypothesis is Anna Olivieri's team's discovery of an ancient T2b3-C151T carrier who died between 1153–983 B.C.E. and was buried in Cagliari, Sardinia[518] since that was the location of the Phoenician colony of Karaly; that sample from Sardinia carried an extra mutation (A9926G) and was given the new subclade name T2b3-a3* by YFull's MTree.

The parent haplogroup T2b3 is found in modern Tunisians, Sardinians, and people in Spain, Portugal, and Russia.

T2b4

Ashkenazim are rarely assigned to the mtDNA haplogroup T2b4. At least one full Ashkenazi with a matriline from Ukraine was fully sequenced into T2b4 by Family Tree DNA. T2b4 is also found among such ethnicities as the Czechs, the Greeks, and the Kazakhs and some people from countries like Denmark, England, Turkey, and Iran. Its Ashkenazic variety, named T2b4-a5 by YFull's MTree, is defined by the mutation A16241G and it is also found in the Catalonia and Castilla-La Mancha regions of modern Spain.[519]

T2b4a

The branch T2b4a is even less common in Ashkenazim than is T2b4. I confirmed that an Ashkenazi who tested with Family Tree DNA whose matriline is Jewish from Belarus was fully sequenced into T2b4a. T2b4a is also found among many non-Jewish people in Denmark and England plus some in Portugal and in ethnic Bulgarians and Ukrainians.

A T2b4a sample from the tenth or eleventh century was found in a Hungarian cemetery.[520] Another medieval T2b4a carrier was found in a chamber grave in a cemetery in Kałdus in north-central Poland.[521]

A daughter lineage, T2b4a1, was carried by a person found in an archaeological excavation at the ASR 13 II Lindegården site in Ribe, Jutland, southern Denmark.[522] Today, T2b4a1 is in modern Denmark, Sweden, and Italy.

T2b16

T2b16 is an uncommon mtDNA haplogroup among Ashkenazim. It was found among Jews in the Russian Empire,[523] including Lithuanian Jews. Outside of Jewish communities, T2b16 exists in today's German, Czech, Slovak, Volga Tatar, and Kazakh ethnicities as well as in Scandinavia and Estonia and among Italians in northern Italy as well as the Campania and Calabria regions of southern Italy. Inside Family Tree DNA's Full Coding Region screen, Ashkenazim are exact matches to some of these members of the German, Czech, Slovak, Scandinavian, and Calabrian populations.[524]

A grave found in Barbing-Irlmauth, Bavaria, southern Germany dating to between 580–620 contained a male who belonged to haplogroup T2b16.[525] A female carrier of T2b16 who was part of the Wielbark culture was buried sometime between the years 80 and 260 C.E. at what is now Kowalewko in Greater Poland[526] and since this culture might be Gothic it could enhance the connection of this haplogroup to Germanic peoples. A person who died between 1500–1320 B.C.E. who was found at the Häffinds archaeological site on the eastern Swedish island of Gotland likewise carried T2b16.[527]

Since some Germanic Langobards (called *Lombardi* in Italian) ruled over Campania from the sixth century onward, it is likely that this explains the presence of T2b16 among Italians in Campania today. I would therefore argue that all evidence combined would support the idea that Ashkenazim could have inherited T2b16 from a German woman rather than an Italian woman.

T2b25

T2b25 is a moderately common mtDNA haplogroup among Ashkenazim. Some of their matrilines stem from Ukraine, Poland, Lithuania, Slovakia, Hungary, and Austria. The Ashkenazim see one person with an ethnic Polish matriline and one person with an ethnic German matriline as their non-Jewish mtDNA T2b25-carrying matches inside Family Tree DNA as of the time of this writing.[528] T2b25 has also been found in matrilines from Italy, Greece,[529] and Bulgaria. That includes a Sicilian from Sciacca, a town in the province of Agrigento, who tested with Family Tree DNA but is not a close match to the Ashkenazim.[530] A daughter branch called T2b25a was present in Roman-era Italy and a branch of that called T2b25a1 is present in non-Jewish Russians[531] and Hungarians. Some people in Spain belong to a branch that YFull's MTree named T2b25d.[532]

T2b25's parent haplogroup, T2b, has a wide distribution across the world, including such countries as Finland, Norway, Poland, Bulgaria, Italy, Tunisia, and the United Arab Emirates and among Armenians and Uyghurs.

The scientists who found the Hungarian with T2b25a1 made a point of noting that it was one of multiple haplogroups that revealed an aspect of commonality between Hungarians and Slavic peoples.[533] Ashkenazim might have acquired T2b25 from a Slavic ancestor, rather than from a southern European ancestor, due to its apparent absence among German Jews.

T2e1b

Ashkenazic Jews sometimes belong to the mtDNA haplogroup T2e1b and some of them belong to its daughter branch T2e1b1. T2e1b has been identified in Ashkenazic matrilines from Germany, Czechia, Hungary, Ukraine, Poland, and Belarus. T2e1b1 is found in Ashkenazic matrilines from Ukraine, Poland, and Belarus.

Felice Bedford's research team discovered T2e1b1 in a Sephardic Jew with a matriline from Romania and T2e1b in a Sephardic matriline from the Netherlands.[534] A T2e1b1 carrier with a matriline from Greece got tested at Family Tree DNA.

YFull's MTree estimates that it is possible that all carriers of T2e1b1 have a most recent common ancestor who lived circa 1200 but their confidence interval allows the possibility that she may have lived as long ago as 450.

The Crypto-Jews in Bragança in northeastern Portugal descend from Sephardic Jews. Inês Nogueiro's research team located T2e1b in a member of Bragança's Crypto-Jewish community.[535]

The parent haplogroup, T2e1, is found in inhabitants of such places as Portugal, Denmark, Norway, and among the Basques. Genetic sequencing of buried Germanic Langobard individuals turned up one case of T2e1 among them.[536]

T2g1a

The mtDNA haplogroup T2g1a, which YFull's MTree estimates formed about 6,300 years ago, is rare among Ashkenazim but has been found in Ashkenazic matrilines from Poland, Belarus, and Latvia. Family Tree DNA assigns them the haplogroup name T2g1a as of 2021. A Lithuanian Jew whose GenBank entry assigned the haplogroup T2g1 has been assigned the more specific haplogroup T2g1a1 by Fernandes' research team.[537] As of the time of this writing, Family Tree DNA's database does not confirm that any Ashkenazim belong to that branch. Inside Family

Tree DNA, an Ashkenazic carrier of T2g1a has Emirati matches in the Full Coding Region screen with a genetic distance of 3.[538] A carrier of this haplogroup from the United Arab Emirates, not necessarily an identical person, was studied by F. A. Al-Jasmi's research team.[539]

A couple of Southern Italians, including at least one Sicilian, who tested with Family Tree DNA belong to T2g1a.[540] T2g1a has also been found among people in Egypt, Yemen, Iran, Turkey, India, and among Armenians, and Palestinian Arabs. Amazingly, T2g1a is even found among the Yakut and Tungusic peoples of Siberia.

In late-medieval times, a female carrier of T2g1a was buried at the cemetery of the Cistercian abbey of St. Mary Graces in East Smithfield, London, England.[541]

A much more ancient female carrier of T2g1a was buried sometime between 4500–3900 B.C.E., during the Chalcolithic era (Copper Age), in Peqi'in Cave in Upper Galilee, Israel.[542]

A Family Tree DNA customer with a matrilineal ancestor who was a Mountain Jew in Derbent, Daghestan, far-southern Russia carries the daughter subclade T2g1a1.[543] There are Family Tree DNA customers with matrilines from Lebanon, Kuwait, Iraq, Italy, Germany, Hungary, and Russia whose displayed haplogroup is T2g1a1.

The ancestral haplogroup, T2g1, shows a distribution pattern that tends to be in the Middle East. T2g1 has been found in Palestinian Arabs, Emiratis, and in Turkey, but also in Pamir highlanders in Central Asia.

The fully sequenced root-level haplogroup T2g is found among Iranian Jews according to Behar's team.[544] However, they made that determination from a sample that later was more precisely placed into the subclade T2g1a1a by YFull's MTree and even more exactly to T2g1a1a1 by Fernandes' team's study.[545] Fernandes' team also classified a person from Iraq and a person from Italy as carriers of T2g1a1a1.[546]

U1b1

Haplogroup U1b1 is about 12,700 years old according to YFull's MTree. So far, Ashkenazic U1b1 carriers only match other Ashkenazim inside Family Tree DNA.[547] These include Ashkenazim with matrilines from Belarus, Lithuania, Poland, Ukraine, Romania, Hungary, Austria, and

Germany. They carry the mutation A3426G that MTree uses to define a branch they call U1b1a1 that formed about 6,900 years ago.

Non-Ashkenazic carriers of U1b1 exist in other databases. U1b1 and its branch U1b1b are found among Armenians, including those from Turkey.[548] U1b1 was found in a Bedouin person from the Negev region of southern Israel.[549] Sardinians also sometimes carry U1b1.[550]

A Roman-era female carrier of U1b1 was buried in Via Paisiello (Necropoli Salaria), Italy in the first or second century of the common era.[551]

Its parent haplogroup, U1b, is found in modern Lebanon. Iraqi Arabs and Iraqi Kurds sometimes carry U1b as well.

All told, U1b1 is likely to have entered the Ashkenazic population from a Middle Eastern ancestor.

U2e1a1

U2e1a1 is a mtDNA haplogroup that has been found in Ashkenazim tracing their matrilines from places like Moldova and Poland. Ashkenazim share not just the same haplogroup but also the same haplotype with Sephardic-descended Crypto-Jews living in Bragança, Portugal. As Nogueiro's team noted, the commonality between the two groups that is seen with U2e1a1 and T2e1b "could represent two possible scenarios: the defining variants could have arisen before the separation between the two Jewish communities; or it may have resulted from a recent introgression of Sephardic lineages into the Ashkenazi gene pool."[552]

Modern non-Ashkenazic and non-Sephardic carriers of U2e1a1 include various Europeans, including but not limited to ethnic Germans, Danes, Belarusians, Ukrainians, Poles, Czechs, Slovaks, and Bulgarians but also ethnic Armenians and people in Iran, Georgia, Turkey, and Saudi Arabia.

Carriers of U2e1a1 from medieval Denmark and Hungary are known. Multiple samples attest to U2e1a1's presence in ancient Europe. One is a male from the Corded Ware culture who was buried in Esperstedt, Germany circa 2500–2050 B.C.E.[553] Another is a female member of the Bell Beaker culture who was found at Osterhofen-Altenmarkt, Bavaria and dated back to 2500–2000 B.C.E.[554] One of the Bronze Age males who died in combat circa 1230–1180 B.C.E. on the Tollense Valley battlefield in Weltzin in northeastern Germany belonged to U2e1a1.[555] Bronze Age Poland, too, had U2e1a1, as demonstrated by the individual with this haplogroup who was part of the Trzciniec cultural circle and placed in a grave at the Koszyce 3 site circa 1423–1233 B.C.E.[556] A Yamnaya male bearing U2e1a1 was buried circa 3350–3200 B.C.E. in a grave at the Pidlisivka archaeological site in Ukraine.[557] A female carrier of U2e1a1 from the Potapovka culture was buried circa 2465–1981 B.C.E. in a kurgan grave at Grachyovka II in Samara Oblast in the middle Volga region of European Russia.[558] U2e1a1 was also found in the early Xiongnu people of Mongolia.

Its daughter subclades are also present in modern European non-Jews. U2e1a1a is found in countries like Finland, Sweden, Denmark, Germany, England, Ireland, and Belgium. U2e1a1b is similarly found in Denmark, Germany, and England but also in the Netherlands. U2e1a1c is found in Germany, Poland, Croatia, Slovenia, Austria, Scotland, and Italy.

U3a1

Haplogroup U3a1 (also called U3a1*) came into the Ashkenazic population from a different ancestor than the one who introduced U3a1a. I confirmed that multiple fully Ashkenazic people were assigned U3a1 by Family Tree DNA as the result of complete sequencing and that at least one U3a1 carrier has an Ashkenazic matriline from Poland.

U3a1 and its parent haplogroup, U3a, both have a presence among the people of Lebanon and Sardinia. U3a1 is also found in Mauritania, Portugal, Germany, Poland, the county of North Yorkshire in England,[559] and the Macedonian region of Greece and has been in both medieval and modern Czechia. U3a is also found in modern Armenia, Georgia, Iran, Egypt, Algeria, and Morocco. U3a1 had a pre-modern presence in places like Spain, Switzerland, and Denmark.[560]

U3a1a

U3a1a is a mtDNA lineage that is found among Ashkenazim with matrilines from Lithuania, Belarus, and Russia's Smolensk region. It has strong European associations. Non-Jewish people in Denmark, Sweden, and Poland are among its modern carriers. Ethnic Russians also sometimes have it. Many of the Ashkenazic U3a1a carriers are exact matches to one ethnic German and one Finn inside Family Tree DNA's Full Coding Region screen. The German has a Catholic matriline descended from Dorothea who was born in the mid-seventeenth century in Bavaria's Lower Franconia district.[561]

Ian Logan, Wim Penninx, and YFull's MTree identified A257G as a special mutation that the German carrier and all Ashkenazic carriers of U3a1a possess. Logan and YFull call their subclade U3a1a3 but that nomenclature is not used by Family Tree DNA.

It is likely that Ashkenazim acquired U3a1a from a German or Polish convert to Judaism. Four ancient samples from Poland were sequenced with this haplogroup. Two of them were females who were found buried in the Wielbark culture's cemetery at Kowalewko, Poland and date to between the years 80 and 260 C.E.[562] Another pair of U3a1a carriers, one female and one male, were found in a different Wielbark culture site, in Masłomęcz, Poland, where they were buried sometime between 200 and 400 C.E.[563]

Its daughter lineage U3a1a1 has been found among non-Jews in Poland, Ireland, and Italy.

U4a3a

U4a3a is an uncommon haplogroup among Ashkenazim. At least some of its Ashkenazic carriers have matrilines from Ukraine[564] and the Transylvanian region of Romania and they have the extra mutation T13659C. U4a3a is also found in a variety of modern non-Jewish Europeans, including, but not limited to, the English, Swiss, Germans, Norwegians, Czechs, and Poles. U4a3a was already present in medieval[565] and ancient[566] England.

Unlike for U4a3a, its parent haplogroup U4a3 has sometimes been found in more southerly lands. For instance, a matriline tracing to an ethnic Armenian woman from Turkey carries U4a3, as determined by complete sequencing.[567] Undifferentiated U4a3 has also been found in modern Italy and Germany.

U4a3's parent haplogroup, U4a, has been found at a variety of ancient sites in southeastern Europe, eastern Europe, and northeastern Europe. One individual carrying U4a was buried between about

5500–5000 B.C.E. in the Yuzhniy Oleniy Ostrov cemetery within Lake Onega in Karelia in northwestern Russia that belonged to a Mesolithic (Middle Stone Age) hunter-gatherer society called the Kunda culture.[568] Another U4a-carrying individual, a male, was a member of a Mesolithic society known as the Iron Gates culture and was buried in what is now Serbia between 8240–7940 B.C.E.[569] Later, U4a was found among inhabitants of Ukraine during the Neolithic era and inhabitants of Bulgaria and Armenia during the Chalcolithic era.

U5a1a2a

Haplogroup U5a1a2a, which formed about 8,400 years ago, is extremely rare in Ashkenazim. I managed to verify that at least one full Ashkenazi who tested with Family Tree DNA is assigned U5a1a2a as the result of complete sequencing. I believe that Ashkenazim would have received this haplogroup from a European proselyte.

Many carriers of U5a1a2a who participate in Family Tree DNA's mtDNA Haplotree list their matrilines from Ireland, Germany, and England, and some of the others list other European countries including Scotland, France, Sweden, Finland, Italy, Belarus, Lithuania, Estonia, Russia, Ukraine, and Belgium. A handful have matrilines from Turkey and India. Scientific papers found U5a1a2a carriers in countries including Serbia, Germany, Denmark, and Poland and in ethnic Russian, Buryat, Uyghur, Armenian, and Iranian people.

A number of pre-modern U5a1a2a carriers have been identified. U5a1a2a was sequenced in a member of the Magyar conquest culture who died between 890–950 and found at Karos-Eperjesszög cemetery number three in Hungary.[570] A female U5a1a2a was found in a grave at the Benzingerode-Heimburg site in Germany and her remains date to between 2287–2041 B.C.E. during the Neolithic epoch.[571] A Bronze Age U5a1a2a male was found at Hastings Hill in northeastern England and he died between 1930–1759 B.C.E.[572] An early U5a1a2a carrier outside of Europe was a female from the Karasuk culture who died in approximately 1414–1261 B.C.E. and was found in Russia's Altai region.[573]

The subclade U5a1a2a1 is likewise found in Ireland, England, Poland, Germany, and Denmark, and a subclade of that called U5a1a2a1a is found in Sweden, France, Switzerland, and Ireland and especially among the

English. Continuing the family's European associations, the sister haplogroup U5a1a2b is found today among Danish and Polish people and in Spain and it existed in ancient Germany, Czechia, Spain, Wales, and Moldova, and among the ancient Scythians. Their ancestral haplogroup U5a1a is in Latvia, Russia, and Scotland.

U5a1b

Haplogroup U5a1b, also called U5a1b*, is found in a low proportion of Ashkenazim. I received confirmation, and later confirmed myself, that this is the complete mitochondrial sequence for certain full Ashkenazim who tested at Family Tree DNA. At least one of them has a documented matriline from Belarus while another's matriline is from Romania.

Family Tree DNA's mtDNA Haplotree lists a large number of carriers of U5a1b who trace their matrilineal ancestries to other European countries, especially England, Germany, Finland, Ireland, and France but also Poland, Norway, Denmark, Scotland, Wales, Spain, Italy, Greece, and others. Most of those carriers are not Jewish. Finland's carriers are confirmed to include ethnic Finns. U5a1b has also been found in non-Jews from Russia.[574]

Outside of Europe, U5a1b is found in ethnic Armenians (including those from Turkey), in Pamir peoples (including Chinese Tajiks who speak the Sarikoli language in Xinjiang), and in Iran. This haplogroup is also found among Kazakhs in China and/or Mongolians who live in China's Inner Mongolia Autonomous Region.[575]

Two pre-modern U5a1b carriers were found in Denmark. One was buried at the rural Faldborg site ("VSM 29F") in Viborg in late medieval times[576] and the other was buried in the urban cemetery of Horsens Klosterkirke ("HOM 1272") in a post-medieval period.[577] U5a1b was identified in one Germanic Langobard who was buried at the Hegykő site in Hungary between the years 500–600.[578]

A Scythian who lived in the North Pontic region in eastern Europe during the first millennium B.C.E. belonged to U5a1b.[579] This haplogroup has a very deep history in the Pontic steppe. Four Neolithic U5a1b carriers were found at the Dereivka cultural site in Ukraine's North Pontic region and all of them were males who died between 5500–4800 B.C.E. and three of them were immediate relatives.[580]

A U5a1b carrier from the Lchashen-Metsamor culture was buried at Dari Glukh in Armenia sometime between 1300–1100 B.C.E.[581] The Zvejnieki burial ground in northern Latvia included a female from the Corded Ware culture who carried U5a1b and died between 3089–2676 B.C.E.[582] Another carrier of U5a1b from the Corded Ware culture was buried in a grave under a kurgan in the village of Hubinek in southeastern Poland between 3025–2898 B.C.E.[583]

U5a1b1

This is an uncommon Ashkenazic mtDNA haplogroup, but I was able to verify that it is held by several full Ashkenazim who had their complete sequences tested by Family Tree DNA. U5a1b1 has also been found in people from countries including, but not limited to, Poland, Germany, the British Isles, Ireland, Finland, Switzerland, Italy, and Portugal. A daughter branch called U5a1b1a is found among many people of England and Ireland. U5a1b1c is found in non-Jews from countries including Scotland, Germany, Sweden, Russia, Poland, Ukraine, and Greece and, as I will explain in the following entry, U5a1b1c2 is also very European. U5a1b1b1 are found in Russia and Belarus, U5a1b1d and U5a1b1e are found in Poland, and U5a1b1h has been found among both pre-modern and modern individuals from Finland. The geographical distributions that we witness for this family of haplogroups suggest that the first Ashkenazic U5a1b1 was a European proselyte.

A Germanic Langobard carrying U5a1b1 was unearthed at an archaeological site.[584]

U5a1b1c2

I confirmed that U5a1b1c2 (a granddaughter subclade of U5a1b1) is the complete mtDNA sequence of at least one fully Ashkenazic person who tested with Family Tree DNA. I and other researchers saw that it is very rare among Ashkenazim. A European proselyte presumably introduced this haplogroup into the Ashkenazic gene pool. Family Tree DNA's mtDNA Haplotree lists U5a1b1c2-carrying customers with matrilines from Poland, Slovakia, Russia, Kazakhstan, Sweden, Scotland, Germany, Lithuania, and Czechia. YFull's MTree identified several descendant subclades of this haplogroup and assigned them unofficial names. One is U5a1b1c2a, which is found in at least one matriline from Poland, perhaps a non-Jewish one. Another is what they call U5a1b1c2a1, which was found in a non-Jewish matriline from Russia.[585] The latter sample was simply called U5a1b1c2 in the original published study.

A carrier of the root level U5a1b1c2* who died in the eighth or ninth century, during the late Avar period, was found in a cemetery in the village of Vörs in Somogy County in western Hungary.[586]

U5a1d2b

U5a1d2b is an extremely rare haplogroup among Ashkenazim. It was not found during earlier comprehensive investigations of Ashkenazic U haplogroups by Costa's team, Wim Penninx, and Jeffrey Wexler. In September 2021, I discovered that a fully Ashkenazic person (per Eurogenes Jtest's oracle) with a Jewish matriline from northeastern Hungary who had just recently tested with Family Tree DNA up to the Full Coding Region level got sequenced by that company into U5a1d2b.

Family Tree DNA's mtDNA Haplotree indicates that their other U5a1d2b-carrying customers have matrilines from Finland, Russia, Sweden, Norway, Poland, Ukraine, and Czechia. Finland is their most common ancestral country of origin listed, followed by Russia and Sweden. Scientific studies confirm that U5a1d2b are found in modern Finland[587] and European Russia west of the Volga,[588] and YFull places them and other modern carriers into newly identified branches of it. U5a1d2b1c is a branch named by YFull that is found in some of their customers from Russia's Ingushetia and Chechnya regions. Other varieties of U5a1d2b have been found in Uralic, North Asian, and East-Central Asian regions, including in a Tatar from the Volga-Ural region,[589] a Bashkir from Bashkortostan,[590] a Pamir highlander,[591] Tubalars from the North Altai region,[592] and Uyghurs.[593] An ethnic Persian carrier from Iran was also identified.[594]

Three U5a1d2b carriers are known to have been buried in Hungary in the tenth and/or eleventh centuries.[595]

U5a1d2b is the assignment for a member of the early Sarmatian culture who was buried at the Pokrovka site in western Siberia's southern Ural region circa 500–200 B.C.E.[596] It is also the haplogroup of a member of the nomadic Aldy-Bel culture who was buried at the Arzhan 2 site in the Tuva region of southern Siberia between 700–600 B.C.E.[597] A female from the Saka culture had this haplogroup and was buried between 500 B.C.E. and the start of the common era in a kurgan in Kyrgyzstan's Tian Shan region.[598] A Neolithic female U5a1d2b carrier who died circa 3630–3360 B.C.E. was found at Tamula, Estonia.[599] A Yamnaya male bearing U5a1d2b was buried in a kurgan grave at the Temrta IV site in southern Russia's Pontic steppe between 3000–2400 B.C.E.[600] Three members of the ancient Afanasievo culture of southern Siberia who died in the 3rd millennium B.C.E. are known to have carried U5a1d2b.[601]

Its sister haplogroup U5a1d2a is found in multiple European populations and its branch U5a1d2a1 is found in northern and eastern Europeans and Buryats.

The parent haplogroup, U5a1d2, is present among the Ket people of Siberia. The grandparent haplogroup, U5a1d, was present in ancient Latvia and ancient Norway and has modern carriers at Family Tree DNA with matrilineal roots in western, northern, and eastern Europe.

Based on the current state of my knowledge of U5a1d2b, I presume that Ashkenazim acquired it either from a Slavic or, perhaps, Khazarian ancestor.

U5a1f1a

U5a1f1a is a mtDNA haplogroup encountered among Ashkenazim with matrilines from Ukraine, Belarus, and Hungary. U5a1f1a is also present among the Adygei people (western Circassians) of the North Caucasus.

A daughter subclade called U5a1f1a1 is found in non-Jewish people with matrilines from Germany, Norway, and England and among Finns and Orkney's Orcadians. Carriers of the ancestral haplogroup U5a1f1* have been found in Switzerland, Italy, and Georgia. The haplogroup U5a1f2 is found among people in Russia and Denmark.

An Early Iron Age male belonging to the Eastern Scythian people who was buried in a grave pit within a mound in central Kazakhstan circa 780–380 B.C.E. was a carrier of haplogroup U5a1f1.[602]

U5a2b2a

The mtDNA haplogroup U5a2b2a is not very common among Ashkenazim. It has been detected in Ashkenazic matrilines from Romania, Ukraine,[603] and Belarus for customers of Family Tree DNA. I thought it could represent the heritage of a Polish convert to Judaism, since its daughter branch U5a2b2a1 is found among non-Jews in Poland and Slovakia[604] and U5a2b2a is not found in German Jews, but as of the time of this writing the only close non-Jewish mtDNA match to the U5a2b2a-carrying Ashkenazim inside Family Tree DNA is an ethnic Swiss person from Signau[605] where most of the people speak German.

Its parent haplogroup, U5a2b2, is found in Europe and Asia. Family Tree DNA customers who carry U5a2b2 include an Austrian and a Ladin person from far-northern Italy. A 23andMe customer who belongs to U5a2b2 is identified in Ian Logan's database as "VahrnTyrol", where Tyrol refers to the region in far-northern Italy and western Austria. Vahrn is a municipality in South Tyrol where the dominant language is German. Many Ladins live in South Tyrol as well. U5a2b2 also exists among Pamir highlanders in Central Asia.

Other branches of U5a2b have Eastern European connections. U5a2b1 is found in Belarus, Russia, Ukraine, Poland, and Czechia, for example. U5a2b1c is found in Poland, Czechia, and Russia and among the Uyghurs. U5a2b1d is found in Belarus.

The grandparent haplogroup U5a2b is found in modern Italy, including among Sicilians, plus Lebanon and Bulgaria and also existed among the ancient Scythians.

U5b1b1-T16192C!

This variety of the mtDNA haplogroup U5b1b1 includes the back mutation T16192C! and is very rarely encountered in Ashkenazim. I confirmed that a fully Ashkenazic customer of Family Tree DNA with a direct maternal line from a Jewish woman from Poland was fully sequenced into U5b1b1-T16192C!.

GenBank lists many samples where the haplogroup assignment is U5b1b1-T16192C without the exclamation point at the end. They include people with matrilines from modern Italy (including Sardinia), Slovenia,

Poland, and Algeria; the individuals with the matrilines from Slovenia and Algeria are listed as having uploaded their data from Family Tree DNA. I saw that Family Tree DNA's mtDNA Haplotree confirms that people carrying U5b1b1-T16192C! with the exclamation point include multiple participants from Slovenia and Poland, two from Italy, one from Algeria, and numerous participants from other European countries, especially England, Ireland, and Germany but also Sweden, Portugal, Spain, Moldova, and more besides.

U5b1b1-T16192C was identified in one Germanic Langobard who was buried at an archaeological site in Hungary.[606] U5b1b1-T16192C is also the assignment for one recovered member of ancient Lithuania's Narva culture.[607] Additionally, two members of the Bronze Age's Trzciniec culture who were buried in graves at Nowa Huta in southern Poland belonged to U5b1b1-T16192C.[608]

Of the aforementioned samples, YFull's MTree places GenBank's participants from Italy and Sardinia in a subclade they call U5b1b1-b* and also places the Langobard and Narva samples in that, whereas GenBank's sample from Poland has the extra mutation C10900T so it is placed under a branch of that that they call U5b1b1-b1. At this time, I do not know where Ashkenazic haplogroup members would fall on their MTree.

A daughter branch of U5b1b1-T16192C! that is called U5b1b1a is found among multiple ethnicities from northern Eurasia including the Norwegians, Finns, Saami, Poles, Mongolians, Buryats, and Yakuts and was in both Finland and Denmark during the Middle Ages. It seems that they descend matrilineally from hunter-gatherers of early northeastern Europe.

By coincidence, a Jew with a matriline from Cochin, India was sequenced into its parent haplogroup, U5b1b.[609]

U5b1e1

Inside Family Tree DNA, I confirmed that U5b1e1 is a mtDNA haplogroup that in rare occasions is found among full Ashkenazim with matrilines from eastern Europe. U5b1e1 has had a wide presence in ancient and modern Europe, primarily. It has a massive presence in modern Finns and Swedish-speaking Finns, an important presence among non-Jewish Poles,[610] and is also found in non-Jews in Sweden, Denmark, Austria,

Slovakia, and Russia, among other places.[611] Surprisingly, U5b1e1 has also been found in two Palestinian Arabs.[612]

U5b1e1 was already present in Poland during the Bronze Age: a Trzciniec cultural circle grave at Żerniki Górne contained a person with this haplogroup who was buried sometime between 1631–1455 B.C.E.[613] Three Viking carriers of U5b1e1 have been found, including two males from Kopparsvik on the Swedish island of Gotland who date to somewhere between 900–1050[614] and a female from Greenland's Western Settlement who died in the year 1080.[615] A U5b1e1 carrier was buried in the urban cemetery of Horsens Klosterkirke ("HOM 1272") in Denmark in a post-medieval period.[616]

It is likely that U5b1e1 traces back to hunter-gatherers who lived in northeastern Europe in earlier times. Its daughter branch U5b1e1a is also found in east-central and eastern Europe, including among Poles, Czechs, and Crimean Tatars and people in Belarus, Ukraine, Germany, Slovakia, and Kosovo. Its parent haplogroup, U5b1e, is found in places including Poland, Russia, Finland, Norway, Germany, Portugal, Spain, and Italy's Umbria region.

U5b2a1a

The European haplogroup U5b2a1a is very rare among Ashkenazim. I received confirmation that U5b2a1a is the terminal subclade for a full Ashkenazi whose maternal line traces to Poland and tested their complete mitochondrial sequence with Family Tree DNA. In the Full Coding Region screen, this Ashkenazi matches only one other Ashkenazi. A more distant match is a German non-Jew whose haplogroup was fully sequenced too but shows up only in the HVR1 screen, not in the Full Coding Region screen, meaning that their genetic distance is greater than 3.[617]

I am aware of a fully Ashkenazic man who is listed simply as U5b2 by Family Tree DNA. However, he did not purchase their Full Mitochondrial Sequence test. If he had, he presumably would also have been assigned to U5b2a1a. His mother and maternal grandmother were Latvian Jews and his great-grandmother in his maternal line was a Lithuanian Jew.

Another carrier of U5b2a1a at Family Tree DNA wrote that his matriline is ethnic Albanian from Kosovo.[618] Family Tree DNA's mtDNA

Haplotree lists other U5b2a1a carriers with matrilines from Germany, the Netherlands, France, England, Scotland, Italy, Spain, and El Salvador and most or all of them must be non-Ashkenazic. Malyarchuk's team found U5b2a1a in members of the ethnic Russian, Polish, Slovak, and Czech populations.[619]

There are many ancient U5b2a1a samples from across Europe and they span from the Mesolithic and Neolithic eras into the Copper Age, Bronze Age, and Iron Age. A grave in Krusza Zamkowa, Poland contained the remains of a young woman who died around 4332–4066 B.C.E. who belonged to haplogroup U5b2a1a.[620] Two females from the Vlasac site in Serbia carried U5b2a1a; one dates to between 7100–5900 B.C.E. and the other to between 6636–6476 B.C.E.[621] A male U5b2a1a hunter-gatherer who died circa 6601–6476 B.C.E. was found at Zvejnieki, Latvia.[622] Male and female carriers of U5b2a1a were found at the Volniensky archaeological site in Ukraine's Pontic steppe region; the male dates to between 5469–5328 B.C.E. while the female died between 6500–4000 B.C.E.[623] A female U5b2a1a who died between 4337–4246 B.C.E. was buried in a grave under Dzhulyunitsa, Bulgaria.[624]

U5b2a1a also existed in ancient western Europe. A female carrier of this haplogroup was buried at the Camino del Molino site in Caravaca, Murcia, Spain between 2920–2340 B.C.E.[625] Two female Bronze Age U5b2a1a carriers who were first-degree relatives of each other were found at Biddenham Loop, Bedfordshire, England.[626]

Its parent haplogroup, U5b2a1, has been found in modern people with matrilines from England, France, Italy, and Sweden.

U5b2a1a's descendant branches are also associated mostly with Europe. Its daughter subclade U5b2a1a-T16311C! is found in matrilines from countries like Finland, Latvia, Italy, Spain, Portugal, Albania, Ukraine, Croatia, Germany, and Scotland but has also been found in an Emirati. Another of its daughter subclades, U5b2a1a2, is found in Bosnians and Basques and in Sweden and Ireland and was found in an ancient Scythian from Ukraine and an ancient Hun from Kazakhstan, and a subclade of that which is called U5b2a1a2a by YFull's MTree is found in Spain, Italy, Finland, Ukraine, and Russia but also in Iran and India. The descendant branch U5b2a1a1 is found in countries like Germany, Norway, Denmark, England, Wales, and Luxembourg and was also in ancient Germany and has been found in an early medieval Langobard in Hungary. A branch of that, U5b2a1a1a, is in countries like Germany, England, Wales, Sweden, and Ireland.

Reaching further back on the haplogroup tree, U5b2a was found in ancient Ireland and Britain. An inhabitant of the large proto-city Çatalhöyük in southern Anatolia (modern Turkey) who died in adolescence sometime between 6450–6380 B.C.E. was sequenced as U5b2.[627]

U6a7a1b

The mtDNA haplogroup U6a7a1b is found in Ashkenazim with matrilines from Germany, Poland, Slovakia, Hungary, Ukraine, and Moldova. It is one of the Ashkenazic haplogroups that connects to Sephardic-descended people and was characterized as a "Sephardic Jewish cluster" by Sécher's genetics team after they confirmed the presence of the haplogroup in a non-Jewish Spaniard, a non-Jewish Mexican, a non-Jewish Cuban from Havana, a person from Algeria, and a person from southern Italy in addition to several Ashkenazic Jews.[628] Subsequently, additional Spanish and Mexican carriers of U6a7a1b were identified by scientists and genetic genealogists.[629] One of the Spanish carriers comes from Jaen in south-central Spain. Another Spanish carrier is from Huelva in Andalusia in southwestern Spain. Still another Spanish carrier declared knowledge of Jewish roots. Some Spanish U6a7a1b carriers show as matches to the Ashkenazim in Family Tree DNA's Full Coding Region screen. The Mexican carriers have matrilines from the state of Zacatecas and genealogists traced them back to Mariana de Quero who was born around 1630.[630] Two Family Tree DNA testers with this haplogroup trace their matrilines to Morocco while another one traces theirs to Portugal.[631] At least some of the Ashkenazic U6a7a1b carriers are an exact match to a Moroccan Berber from Fez on the Full Coding Region screen.

Another discovery after Sécher's team's work came when Pierre Zalloua's genetics team sampled a carrier of U6a7a1b from Lebanon.[632] But the origins of U6a7a1b, and the U6a haplogroup family on the whole, are to be found in North Africa rather than the Levant, even though U6 probably originated in West Asia. It is significant that its parent haplogroup, U6a7a1*, is found today among Moroccans, including Riffian Berbers from northern Morocco, and in people from Mauritania, Tunisia, Libya, and the Jijel province in northeastern Algeria. U6a7a1* also existed among the medieval indigenous Guanche people who lived on the Canary Islands and were closely related to the Berber people. A Guanche sample found at the El Agujero archaeological site on Gran Canaria dated to between 1030–1440 belonged to this haplogroup.[633] U6a7a1* has also been sequenced in people with matrilines from Spain, Portugal, western France, Poland, Hungary, Sweden, Scotland, and England and in a person from Israel whose ethnic and religious backgrounds are unknown to me at this time. An entirely non-Jewish branch of U6a7a1* that is called U6a7a1a is found in France, as is its subclade U6a7a1a1.

At a deeper root, haplogroup U6a7a* was sequenced in a female from Sardinia who died in the second century, during the Roman Empire.[634]

Incredibly old examples of U6a7a's parent haplogroup, U6a7, were found in Taforalt cave in northeastern Morocco's Berkane Province.[635] They date from the Later Stone Age. One is from approximately 12,580 B.C.E. while the other is from approximately 12,485 B.C.E. Another U6a7 carrier was found at the Early Neolithic archaeological site Ifri n'Amr Ou Moussa in Aït Siberne in western Morocco's Khémisset Province and dates from approximately 5130 B.C.E.[636] These samples reinforce the U6a family's deep connection to North Africa.

Multiple other branches of U6a7 are found in modern North Africans: U6a7a2 and U6a7d in Morocco, U6a7b1 in Algeria (but also Spain), and U6a7c1 in Tunisia and Morocco (but also Portugal).

The first Jewish U6a7a1b was a Berber who could have lived soon before the initial split of the Ashkenazic and Sephardic populations if the Ashkenazic carriers did not inherit this from a Sephardic woman who migrated to central or eastern Europe after 1492.

U7a5

The mtDNA haplogroup U7a5 is only called U7a at Family Tree DNA as of 2015–2022 which no longer assigns anybody to U7a5 according to their mtDNA Haplotree but it had been called U7a5 there as of 2012 and was also called U7a5 in Marta Costa's team's 2013 study and Hovhannes Sahakyan's team's 2017 study for the Ashkenazic carriers. Family Tree DNA listed the required mutations for U7a5 as 573.XC, G9300A, A13966G, C14245T, G14869c, C16291T, and T16304C.[637] It is found among Ashkenazim with matrilines from Germany, Austria, France, the Netherlands, Poland, Ukraine, Hungary, Romania, Lithuania, and Latvia. It appears to be an inheritance from the ancient Israelites.

U7a5 is also found among modern Jordanians, Anatolian Turks, Armenians from Armenia and Artsakh, and Iranians but also among non-Jewish Germans from Lower Saxony.[638] Costa's research team found an Iranian Jew carrying U7a5 and remarked that U7a5 "has a likely origin in the Near East or South Caucasus."[639]

This haplogroup was also found in ancient Lebanon; the retrieved sample is assigned U7a in GenBank[640] but more precisely to U7a5 in an earlier iteration of YFull's MTree because it includes the mutation G9300A that defines U7a5 according to what was scientists' accepted definition for that haplogroup.

YFull's MTree currently would place all Ashkenazim into U7a5 but places the modern Turkish and Armenian samples and the ancient Lebanon sample into a newly defined parent of U7a5 that they call "U7a-a" as of 2021. They also place a Saudi Arabian customer of theirs into "U7a-a". They previously agreed with the scientists that U7a5 was the haplogroup for the Turk, the Armenian, and the Lebanon sample but used to call the Ashkenazic branch U7a5a as of 2019. They differentiate between the Ashkenazic and non-Ashkenazic branches because whereas the non-Ashkenazim do carry G9300A like the Ashkenazim also do, they lack the other five branch-defining mutations like A13966G and C16291T that the Ashkenazim have.

The parent haplogroup, U7a, is widespread in the West Asia, Southwest Asia, Central Asia, and South Asia with many Iranians and Saudi Arabians being carriers as well as some Tajiks, Uzbeks, Uyghurs, Azeris, Assyrians, Parsis of India, Kashmiris, Pathans, Veddas of Sri Lanka, Bahrainis, Kuwaitis, and Palestinian Arabs. Russian and Belarusian cases of U7a are also documented. Many subclades, including U7a7a, U7a7a1, U7a8a, and U7a9, have presences in India. Various subclades of U7a are also found among Pakistani, Turkish, Tajik, Uzbek, Iraqi, Greek, and Armenian people, among others.

Hovhannes Sahakyan's team sequenced U7a18 in an Iranian Jew.[641] U7a3a appears to be an Iraqi Jewish lineage; it was found in two members of a Jewish family originating from Baghdad that tested with Family Tree DNA.

The Samaritans are a branch of the Israelite people and have retained Israelite DNA to a large degree and some of them carry the mtDNA haplogroup U7a1.[642] U7a1 and the subclade U7a1a1a are also found in Iran, and U7a1a1, U7a1a1a, and U7a1a2 are carried by some people in India.

U8b1b1

The mtDNA haplogroup U8b1b1 is very rare among Ashkenazim. Among my Family Tree DNA matches, I independently verified that it is carried by at least two full Ashkenazim. U8b1b1 is also found among Armenians,[643] people in Spain,[644] and modern Umbrians living in central Italy and already had a presence among ancient Umbrians: a pre-Roman Umbri person who died around the seventh-fourth centuries B.C.E. was found at the necropolis of Plestia in East Umbria.[645]

Geneticists have also sequenced U8b1b1 from many pre-modern European people. It is noteworthy that three U8b1b1 carriers were found in an Iron Age necropolis at the archaeological site of Polizzello in Sicily.[646] An earlier U8b1b1 carrier, who died sometime between 2014–1781 B.C.E. during the Early Bronze Age, was found at Sicily's Necropoli Castellucciana.[647] Even older than that was the U8b1b1 carrier retrieved from Sicily's Buffa Cave who died circa 2287–2044 B.C.E.[648] U8b1b1 is also the assignment for a female found at Sardinia's necropolis of Su Crucifissu Mannu who died circa 3626–3374 B.C.E. during the Early Copper Age.[649]

Other ancient U8b1b1 samples come from outside of Italy. A U8b1b1-bearing male from the Neolithic-era Starčevo culture was found at Beli Manastir-Popova Zemlja, Croatia and he is estimated to have died between 5837–5659 B.C.E.[650] Two U8b1b1 carriers from the Bell Beaker culture, dating to between 2500–2000 B.C.E., were found at Radovesice, Czechia.[651] Another U8b1b1 sample from Czechia is a Neolithic-era person who died circa 4488–4368 B.C.E. and was found in Prague.[652] There is also a U8b1b1 sample from Switzerland dating to between 4766–4598 B.C.E.[653] U8b1b1 also existed in medieval Hungary.[654]

There are also ancient U8b1b1 samples from outside of Europe. A Neolithic male dating back to around 6500–6200 B.C.E. was found at the village of Barcın in northwestern Anatolia in Turkey and he was sequenced as U8b1b1.[655] A Ptolemaic period female buried circa 344–169

B.C.E. in a tomb at the site of the ancient Egyptian city of Abusir el-Meleq belonged to U8b1b1.[656]

Might U8b1b1 in Ashkenazim represent the input of an Italian woman who converted to Judaism? Its long-standing presence in Italian lands is suggestive of that possibility.

Its parent haplogroup, U8b1b, is found in Sardinians and Uyghurs.

V1a1

V1a1 is a mtDNA haplogroup of European origin that has been confirmed to exist in the Ashkenazic population, including those with matrilines from Lithuania and Ukraine. V1a1 is common among Finnish people and has also been found among the Irish and people with non-Jewish matrilines from Denmark, Russia, Poland, Czechia, Serbia, and Greece. The mtDNA matches to the Ashkenazim inside Family Tree DNA are mostly West Slavs, East Slavs, and Scandinavians but there are also a couple of German non-Jews from Saxony in eastern Germany, whose inhabitants have considerable Slavic admixture.

A V1a1 carrier was buried in late medieval times in the "HOM 1649" urban cemetery in Ole Wormsgade, Horsens, Denmark.[657] A male Viking found in Oxford, England whose remains date to between 880–1000 belonged to V1a1.[658]

Daughter and granddaughter subclades of V1a1 likewise have European distributions but are not shared by Jews. As some examples, V1a1d, V1a1e, and V1a1f have been found in Poland, V1a1c and V1a1g have been found in Finland, and V1a1b1, V1a1f, V1a1h, V1a1h1, and V1a1i have been found in Russia. There are also branches of V1a1 that are found in Denmark, Sweden, Germany, Portugal, Italy, Bulgaria, and Greece. A rare non-European carrier of one of these subclades is a V1a1h1 Pamir highlander carrier from Central Asia. A Bronze Age carrier of V1a1b from the Trzciniec cultural circle whose remains date to between 1427–1277 B.C.E. was found in a grave under what is now the village of Dacharzów in south-central Poland.[659]

Its parent haplogroup, V1a, sometimes called V1a*, is also very European. A tenth-century carrier of V1a* was found at the Capidava necropolis in the province of Dobruja in southeastern Romania and the haplogroup assignment was confirmed by high-throughput sequencing.[660] Other pre-modern V1a* samples were found in Germany and Denmark.[661]

It is quite possible that the first Jewish V1a1 was a Pole who converted.

V7a

The mtDNA haplogroup V7a has been found among a considerable minority of Ashkenazim, including those with matrilines from Poland (especially), Ukraine, Belarus, Latvia, Lithuania, Slovakia, Romania, Moldova, Hungary, Austria, and (to a limited extent) Germany. It has not been located among non-Ashkenazic Jews.

Among non-Jews, V7a is mostly found among European populations. European Christian V7a carriers include ethnic Swedes, Italians, and Saami and members of different Slavic ethnicities, such as Russians and Slovaks, among others. Outside of Europe, V7a has been found in two places in Algeria, raising the possibility that a European slave carrying this haplogroup had been sent to North Africa.

The known branches of V7a are also found in Europe. V7a1 carriers live in Sweden, Finland, Karelia, Denmark, and Norway, and V7a3 carriers live in Sweden, Finland, and Russia. V7a2 has a more southern distribution with carriers in France, Germany, Czechia, and Romania.

YFull's MTree includes more precise information on mutations that form different branches of V7a and has come up with novel names for branches that they've newly identified. A Lithuanian Jewish matriline has been placed specifically into the subclade V7a2c1b*, formerly called V7a7a2* and defined by the presence of the mutation A12753G, as have been somebody with a matriline from Russia who participated in a project by Behar's team[662] and a person from Denmark from a published study where only some of the samples came from ethnic Danes.[663] MTree named a subclade of this V7a2c1b1, formerly V7a7a2a, which is defined by the mutation A3395G, and it is also found in Ashkenazim. V7a2c1b's sister subclade, V7a2c1a, formerly V7a7a1, is defined by the mutation G6734A and is found in non-Jewish Poles.[664] As of the time of this writing, MTree estimates that V7a2c1, formerly V7a7a, the common ancestor of the Ashkenazim and the aforementioned Poles, emerged around 1500 and that the first V7a2c1b carrier lived about 150 years after that.

Their ancestral haplogroup V7a2c*, formerly V7a7*, has been found in a non-Jewish Russian from the Pskov region.[665]

Five carriers of V7a have been found in medieval commoner graves in Hungary: two at the Homokmégy-Székes cemetery that was in use from the early tenth through early eleventh centuries, one at the Ibrány Esbóhalom cemetery of the tenth and eleventh centuries, one at the Magyarhomorog-Kónyadomb cemetery of the tenth-twelfth centuries, and one at the Sárrétudvari-Hízóföld cemetery of the tenth century.[666]

V7a likely represents a Slavic contributor to the Ashkenazic population. That woman was probably Polish, but Leo Cooper thinks it is also possible that she was a member of a different West Slavic group. Either way, it would be a rare example of a Slavic haplogroup that some modern German Jews share with Eastern Ashkenazim, but Cooper and I do not think early Jews of the Rhineland had it. It is not impossible that early Jews in Prague did.

V7b

A moderate proportion of Ashkenazim carry the mtDNA haplogroup V7b. Some of them have matrilines from Lithuania, Ukraine, and Hungary. V7b is not found in German Jews.

Aside from themselves, Ashkenazic V7b carriers who tested with Family Tree DNA most closely match an ethnic Czech person whose matriline comes from northern Bohemia near Germany. In the Full Coding Region screen, at least some of them match the Czech at a genetic distance of 2. Those same Ashkenazim have a genetic distance of 3 from a Montenegrin, an ethnic Ukrainian from Odessa, and a person with a matriline from France with French Christian first and last names.[667] A Family Tree DNA tester with a matriline from Lithuania tracing back to a woman born around 1830 who had ethnic Lithuanian first and last names also carries V7b.[668] YFull's MTree lists a V7b carrier with an ethnically Polish matriline from Mazovia Province in east-central Poland.

A man from the tenth-century Viking Age gravesite in Bogøvej, Denmark was found to carry V7b.[669]

V7b appears to reveal the story of a Slavic woman who joined the Ashkenazic people.

V15

Among Ashkenazim, V15 is a very rare mtDNA haplogroup. I received confirmation that it existed in at least one Ashkenazic matriline in Ukraine in the nineteenth century and later confirmed it in an Ashkenazic matriline from nineteenth-century Poland. V15 has deep roots in Europe. Family Tree DNA's customers who carry V15 include people with matrilines from England, France (among non-Jews),[670] Italy, the Netherlands, and Armenia (among ethnic Armenians). There are also other V15 carriers there who declared their matrilines to originate in Poland and Hungary; it is not known to me whether those are Ashkenazic, but I suspect that some of them are. The closest mtDNA matches my Ashkenazic correspondent who carries V15 has are the people with the heritage from Poland, Hungary, and Armenia.

A daughter subclade of V15 called V15a is solidly European with numerous non-Jewish carriers with matrilines from Ireland, Scotland, and England along with some with matrilines from Sweden, Norway, Denmark, and Germany.

Unless Ashkenazim acquired V15 from a woman whose ancestors came from the South Caucasus, they probably got it from a proselyte in east-central Europe.

V18a

V18a is a very rare mtDNA haplogroup among Ashkenazim. I confirmed that at least two fully Ashkenazic customers of Family Tree DNA belong to V18a. One of them has been living in Hungary in the twenty-first century and the other has a documented Jewish matriline from Slovakia.

This haplogroup is otherwise totally European. Multiple non-Jewish carriers of V18a have been found in Denmark and an ethnic Czech case of V18a is also known. A V18a sample from Poland was collected by Katarzyna Skonieczna's genetics team.[671]

YFull's MTree assigns some Danish V18a carriers to subclades they call V18a1, V18a1a, and V18a2b (formerly V18a3). Additionally, MTree identified a branch of V18a that they call V18a2 that is found in Sweden and Poland plus a branch they call V18a1a that is in Poland (represented by the aforementioned sample from Skonieczna's team that Ian Logan confirmed is in the V18a family).

A person from southeastern Spain is at the root of V18a*[672] and lacks the extra mutation G7762A that is common in V18a matrilines from Denmark and Poland.

V18a carriers in Family Tree DNA include two people with non-Jewish matrilines from the Netherlands: one from the Limburg province in the southeast and the other from a coastal town in South Holland.[673] Additional Family Tree DNA customers with V18a trace their matrilines to women from Switzerland, Austria, Germany, England, Scotland, and Ireland, all or almost all non-Jewish.

The data we have hint at a likely Polish or German source population for Ashkenazic V18a lineages.

V18a's parent haplogroup, V18, was sequenced in an Iron Age Roman female from Italy.[674] Modern V18 carriers who tested with Family Tree DNA have matrilines from Italy, Austria, Slovenia, and Slovakia.

W1h

Ashkenazim tracing their matrilines to Belarus, Ukraine, and Russia proper and apparently Lithuania as well are among the carriers of haplogroup W1h inside Family Tree DNA. Other W1h-carrying customers of that company trace their matrilines to Italy, Norway, Sweden, England, Croatia, and Finland. W1h is about 6,000 years old according to YFull's MTree.

Ashkenazic carriers have the mutation A4917G that MTree uses to newly identify a daughter branch they call W1h3 that formed about 4,900 years ago. But some Italian carriers have the mutation G513C instead, which defines the branch W1h1 that likewise formed about 4,900 years ago. Italian W1h1 carriers from Calabria[675] and Tuscany[676] were found by scientific teams. Many other W1h1 carriers with matrilines from Italy tested with Family Tree DNA.

In theory, Ashkenazim could have inherited W1h from an Italian woman, but W1h's apparent absence in German Jews suggests that a Slavic ancestor introduced it.

Haplogroup W1 is found today among people whose matrilines trace to a wide variety of European countries ranging from Albania to Portugal to Italy to Ireland to Finland, among many others, and also to one from Pakistan. Three pre-modern W1 carriers were found in Poland.[677]

W3a1a1

Some Ashkenazim with matrilines from Ukraine and Poland carry the mtDNA haplogroup W3a1a1.[678] Defined by the presence of the mutation G3421A, this is a descendant of haplogroup W3a1a. At least some of the Ashkenazic W3a1a1 carriers who tested with Family Tree DNA match an ethnic Pole with a genetic distance of 1 in the Full Coding Region screen and are more distant matches to a few more Poles and one Czech who display only in the HVR1 and HVR2 screens.[679]

YFull's MTree further divides W3a1a1 into W3a1a1a and W3a1a1b. W3a1a1b, with the extra mutation C10715T, is the branch with the Ukrainian Jewish matrilines. YFull found W3a1a1a in two Silesians from the Silesia region of southwestern Poland. Silesians are considered a branch of the Poles. As a result, we can deduce that a Polish woman was responsible for introducing W3a1a1 into the Ashkenazic gene pool. This is more likely than a Ukrainian or Czech woman.

W3a1a has been found among non-Jewish people from Ukraine, Belarus, Poland, Germany, Sweden, Finland, Italy, England, and India. A different branch of W3a1a, called W3a1a2, has been found among the Volga Tatar population of Russia[680] in addition to non-Jews from England, Scotland, and Ireland. W3a1a3 is likewise found in England and Ireland.

The tenth-century Hungarian cemetery Sárrétudvari-Hízóföld contained one male who belonged to W3a1a.[681]

W3a1a likewise had a presence in eastern Europe in ancient times. Significantly, W3a1a was found in two members of the Yamnaya culture. One of the Yamnayan W3a1a carriers was buried in southwestern Russia, in one of the kurgan graves at the Lopatino II site near the Sok River in the Samara region, and this person died circa 3300–2700 B.C.E.[682] The other Yamnayan W3a1a carrier was buried at the Porohy archaeological site in Ukraine and dates to between 2882–2698 B.C.E.[683]

W3a1a's parent haplogroup, W3a1, has a modern presence among non-Jews from countries including France, Italy, Turkey, and India, and the subclade W3a1b is also found in India. In ancient times, W3a1 was the haplogroup of a male Yamnayan whose remains were found at the

Prydnistryanske archaeological site in Ukraine and date to between 3023–2911 B.C.E. as well as of two remains found in Germany: a male from the Únětice culture who was buried between 2134–1939 B.C.E. at what is now Esperstedt and a male from the Bell Beaker culture who was buried between 2350–2250 B.C.E. at Künzing-Bruck in Bavaria.[684] A W3a1 carrier was buried in Kazakhstan in roughly 600 B.C.E. while two other W3a1 carriers were buried in Kazakhstan around the year 300 C.E.[685]

W3b1

W3b1 is occasionally encountered among Ashkenazim, including those with matrilines from Poland, Ukraine, Belarus, and Lithuania.[686] Ashkenazim with this haplogroup who tested with Family Tree DNA match a person whose listed matrilineal ancestor was an ethnic Austrian from Graz in southern Austria.[687] Based on this, I propose that the first Ashkenazic W3b1 was a central European convert to Judaism.

Its parent haplogroup, W3b, is widely distributed across Eurasia, being found among Armenians, Assyrians, Tajiks, the Burusho people of Pakistan, and people residing in Iran, Turkey, Bulgaria, Serbia, and Cambodia. W3b's presence among Armenians is very old as shown by the discovery of a carrier dating to between 1400 and 1200 B.C.E. who was found at a necropolis site under the Armenian village of Nerkin Getashen.[688]

X2b7

Ashkenazim with matrilines from Romania, Moldova, Ukraine, Poland, and Belarus are among the carriers of the mtDNA haplogroup X2b7.[689] One of the only other populations among which X2b7 has been found are matrilineal descendants of French Catholic people, including of the sixteenth-century woman Jeanne Beaudinet,[690] who is believed by some genealogists to have been born in Piney in north-central France. Beaudinet's descendants moved to the North American colony of Acadia,[691] becoming known as Acadians, and some subsequently moved to southwestern Louisiana to form part of its Cajun community. X2b7 is also found among Basque people from northern Spain[692] and was sampled in a person from Valencia in eastern Spain.[693] The Basques and the Valencian share the mutations T14110C and A10634G with the Ashkenazim; the latter is a mutation that YFull's MTree uses to define an unofficial subclade it calls X2b7a. Beaudinet's descendants, on the other hand, carry not T14110C or A10634G but C3942T instead, which YFull and the "X mtDNA Haplogroup" project at Family Tree DNA use to define an unofficial subclade called X2b7b.

X2b7's parent haplogroup, X2b, as well as a distinct branch called X2b4 are similarly found among French people. Originally, I had thought that X2b7 is connected to the Jewish community of Tsarfat that existed in northern France until their expulsions in the fourteenth century,[694] but at least as of the time of this writing, people from Spain are genetically closer to the Ashkenazim with X2b7 than French people are.

X2e2a

X2e2a is quite uncommon among Ashkenazim. At least one of the Ashkenazic members of this haplogroup at Family Tree DNA has a documented matriline from Ukraine. This lineage is typically found in West Asia and the Caucasus region, including among ethnicities in Turkey, Georgia, and Chechnya, and it appears to have originated in West Asia.[695] It has also been found in people from England (including Cornwall) and Greece and among the Teleut, Pamiri, and Basque peoples. The Druze people of the Levant also have X2e2a present among them.[696]

A young adult man bearing X2e2a was entombed at the Vagnari necropolis in southeastern Italy sometime between the first-fourth centuries C.E., during the Roman epoch.[697] A male carrier of X2e2a who died circa 1883–1776 B.C.E. was found at the ancient city-state of Alalakh in Hatay Province in far-southern Turkey.[698]

The Tubalar and Altaian (Altai-Kizhi) peoples who live in the Altai region of central Asia, plus some people who live in the Baikal region of southern Siberia, carry a branch of X2e2a called X2e2a1.[699]

Another branch of X2e2a is X2e2a2, found among Yemenis,[700] Armenians,[701] Mingrelians from Georgia, and people from Azerbaijan and the North Caucasus.

Although X2e was present among the medieval Khazars, there is no reason to believe that Ashkenazim inherited X2e2a (a specific descendant branch of X2e) from a Khazarian ancestor. The Ashkenazic X2e2a is also quite distinct from the haplogroup X2e1a1a that is found among Libyan Jews and Tunisian Jews.[702]

Miscellany

H10*, J1c2c2a, T1a5, T2b-T16362C, T2f1, U4a*, and/or V21 might also be legitimate (albeit rare) Ashkenazic mtDNA haplogroups inside Family Tree DNA but I could not verify those beyond doubt as of the time of this writing.

Other haplogroups listed as Ashkenazic or potentially Ashkenazic in some databases, including B2b3a, H1q, H3k1a, H44b, J1b1a3, J1b2, K1a11, L2a1f, R0a2c, and U3b1b, among others, are entirely lacking in substantiation or disproven by family trees and cannot be considered genuinely Ashkenazic but, instead, reflect recent intermarriages with non-Ashkenazic women.

Leo Cooper and I saw that the European mtDNA haplogroup H39 is occasionally assigned to full Ashkenazim who test with 23andMe. According to Ian Logan and YFull's MTree, H39 is defined by the mutation A16299G and there are 23andMe H39 carriers as well as Family Tree DNA H39 carriers who carry that same mutation in Logan's database. Logan's Family Tree DNA H39 carriers uploaded their sequences to GenBank in 2008–2017. The two who uploaded in 2008 and 2010 listed the unspecific haplogroup H as their assignment inside Family Tree DNA at that time while the Swede who uploaded in 2017 indicated that Family Tree DNA assigned them to H39. However, for some reason, Jeffrey Wexler, Wim Penninx, and I have not found any full Ashkenazim carrying H39 inside Family Tree DNA, so they presumably are assigned to a different haplogroup name there (and therefore in my encyclopedia) as of 2018–2022, but I do not know which one 23andMe's H39 equates to.

Chapter 3: Non-Ashkenazic Haplogroups in Populations Related to Ashkenazim

Non-Ashkenazic Jewish lineages that are not shared by Ashkenazic Jews

Although some maternal haplogroups (including H1bo, H4a1a3a, H6a1a1a, H13a1a1, H25, HV1b2, J2b1e, K1a1b1a, K1a4a, K1a9, N1b1b1, R0a2m, R0a4, T1a1j, T1b3, T2e1b, and U7a5) bridge Ashkenazim with other Jewish communities, including other Western Jews and Mizrahi Jews, there are a considerable number of mtDNA haplogroups that are found in other Jews but not in Ashkenazim. Some of those lineages are unique to particular Jewish communities due to intermarriage with local converted women. Other lineages just happened to not get acquired or retained by the Ashkenazic community due to different migration histories or the murders of many German Jewish women in medieval times.

Moroccan Jews are a particularly diverse population with mixed Judean, European, and Berber ancestry. The geneticist Razib Khan confirmed their partial Berber descent using autosomal DNA admixture modeling. They sometimes carry the Western European mtDNA haplogroups I4a1a, K1a2a, K1a3a1a, and K1c2, the originally Middle Eastern haplogroup HV1a2 (also found among the Palestinian Arabs, Syrians, Assyrians, Qataris, and ancient Armenians but also in Tunisia and some countries in southern and western Europe), the likely Italian haplogroup HV1c, the European haplogroup H1e8 (also found in Spain), the European haplogroup X2d2 (also found in Sardinia and Sweden), the African haplogroups L1b1a and L2a1p, and the variegated haplogroup HV0d (also found in Spain, Italy, England, Denmark, and India). None of these are found among Ashkenazic Jews. Many Moroccan Jews who tested with 23andMe, which does not do complete sequencing, were found to have the mutation A14053G and are assigned by them to the European haplogroup H1o, which some Ashkenazim (including at least one German Jew) who tested with 23andMe also got assigned, but I do not know which H1 family haplogroup in my encyclopedia (which is based on Family Tree DNA's newer nomenclatures in complete sequences) H1o equates to.

HV1c is shared by many Jews from Algeria. Some other Algerian Jews have the non-Ashkenazic haplogroups K2a5 and H4a1b, according to 23andMe, or branches of those haplogroups that could be determined by complete mitochondrial sequencing. Varieties of K2a5 are also found in Portugal, northern Europe, Iran, and South Asia. Varieties of H4a1b are also found in Italy.

The two most frequent mtDNA haplogroups among both Libyan Jews and Tunisian Jews are X2e1a1a and H30. Neither of them is found among Ashkenazic Jews. The Tunisian Jewish haplogroup H3c1 is not found in Ashkenazim but is found in non-Jewish Europeans, including in Italy, France, and Ireland.

Sephardic Jews from Greece, Turkey, and Bulgaria sometimes carry the mtDNA haplogroup T2e1a1a, formerly called T2e5, which is absent from the Ashkenazic gene pool. They share T2e1a1a with Catholic descendants of Sephardim in Mexico and crypto-Jewish descendants of Sephardim in Portugal.[703] H1p, J1b2 and H14b are other mtDNA haplogroups that Sephardic Jews from Turkey have but Ashkenazic Jews do not. H1p was probably originally European and has also been found in a Pole. H14b has also been found in Assyrians, an Armenian, a Turk, Emiratis, and people from Tuscany and Ibiza. The mtDNA haplogroup H1v has been identified as a lineage of Sephardic Jews from Italy, but it is not shared by Ashkenazim either, though it is shared with Mozabite Berbers from Algeria.

Romaniote Jews from Greece descend from Judeans who intermarried with southern Europeans. Romaniote Jews have some mtDNA haplogroups that are not found among Ashkenazic Jews, including U5b and U6a3.[704] U6a3 is also found in Bulgarian Jews[705] and Palestinian Arabs and was in ancient Egypt and ancient Morocco. The subclade U6a3a is likewise present among Bulgarian Jews[706] and has been found in Spain.

Syrian Jews and Egyptian Jews share the mtDNA haplogroup T2c1a1. It is never found in Ashkenazic Jews. T2c1a1 is also found among people in Iraq and Turkey. The Syrian Jewish haplogroup J1b1a3 is not found in Ashkenazim either. Other J1b1a3 carriers include Palestinian Arabs, Armenians (including from Turkey), Sardinians, and Czechs. A Syrian Jew who tested with 23andMe got assigned to the West Asian mtDNA haplogroup H20a, which Behar's team had previously found in Turkish Jews.[707] H20a is also found in Armenians while H20a1 is in Iran and the Lebanese and H20a1a is in Turkey and the United Arab Emirates.

Iranian Jews (Persian Jews) occasionally do share U7a5 with Ashkenazim but also have some other branches of U7a. These include U7a4b1a and U7a18. Their haplogroups H14a and J1b1b1a are not found in Ashkenazim either, but H14a is also found in Iraqi Jews from Baghdad according to Wim Penninx. Iranian Jews descend from a combination of Judean, Mesopotamian, and Persian ancestors.

Mountain Jews of the Caucasus, a branch of Mizrahi Jews that is closely related to Iranian Jews and directly descended from them, sometimes carry the mtDNA haplogroups J2b1f and U7a4a1, unlike Ashkenazic Jews. J2b1f is also found among people in Syria, Turkey, and Armenia, while U7a4a1 is also found among Armenians and people in Iran.

The majority of Georgian Jews, another branch of Mizrahim in the South Caucasus, carry the mtDNA haplogroup HV1a1a1,[708] which is absent from the Ashkenazic population. That probably represents a local convert to Judaism because HV1a1a is also found in non-Jewish Georgians, Armenians, and Assyrians.

Iraqi Jews harbor a number of maternal lineages that no Ashkenazi possesses. According to Behar's team's 2008 study, these include the Middle Eastern haplogroups H13a2b, T2c1, U3b1a, and W1d.[709] In 2021, Wim Penninx found that a branch of H13a2b called H13a2b1 is present among Iraqi Jews and Mountain Jews who tested their complete sequences with Family Tree DNA. H13a2b1 is also found in Armenians and in Jordan, Kuwait, Iran, and Italy.

The Bukharan (Bukharian) Jews of Uzbekistan, Tajikistan, Afghanistan, and Kyrgyzstan largely descend from Iranian Jews. Their genetics have remained strongly Middle Eastern to this day. Based on a large data set of Bukharan Jews' autosomal DNA matches at 23andMe, which fully sequences some mtDNA haplogroups but only partially sequences some others and did not keep up with Mannis van Oven's latest nomenclatures, Leo Cooper inferred that the non-Ashkenazic mtDNA haplogroups J2b1a2, H14, H28, I4, F1b1, and F1b1a1 are particularly prominent in the Bukharan Jewish population.[710] Their variety of H14 is potentially identical to the H14a of Iranian Jews. It is not certain that 23andMe's assignments F1b1 and F1b1a1 are up to date. F1b1 is also found for instance in Armenians, Azeris, Uyghurs, Kyrgyz, and in Iran and Tibet. F1b1a1 (evidently found in 5.7 percent of Bukharan Jews) would be peculiar because it has otherwise been confirmed to be in Japanese and Korean people, who are completely East Asian and live far from Central Asia, and Bukharan Jews score only 0.8 to 0.9 percent East Asian in autosomal tests. U1a1, another haplogroup never encountered in Ashkenazim, appears to have a moderate presence among Bukharan Jews and is also found in Assyrians, Sorani-speaking Kurds in Iraq, Zazas in eastern Turkey, Circassians, and Algerian Jews. T2c1 appears to be another Bukharan Jewish haplogroup and, as noted above, it is also found in Iraqi Jews. Another seems to be U6a2, which has varieties that are found in Emiratis, Kuwaitis, Assyrians, and in Armenia, Italy, and Ethiopia. Other haplogroups that Bukharan Jews appear to possess include varieties of R0, T1, X2, and N but because 23andMe probably did not fully sequence these it is not possible to know whether or not they match Ashkenazic subclades. Apart from Cooper's research, I learned that a person with a Bukharan Jewish matriline from Uzbekistan belongs to H29, according to 23andMe. As of the time of this writing, I do not have any complete mitochondrial sequences for Bukharan Jews who may have tested at Family Tree DNA and there are no published studies on them.

The Yemenite Jews are a highly distinctive Jewish population. A minority of their ancestry traces back to the Judeans, but the majority of their ancestry stems from southern Arab converts during the Himyarite period in Yemen's history when the royal house and army converted to Judaism.[711] As a result, Yemenite Jews have many mtDNA haplogroups that are absent in Ashkenazim, including R0a1c (of Middle Eastern origin), L3x1a (branches of which are also found in Saudi Arabia and East Africa), and R2a (also common in Saudi Arabia).

The Cochin Jews from South India, also known as Malabari Jews, are mostly descended from Indian people, particularly women, who converted to Judaism although they retain a minority of Middle Eastern autosomal DNA that is similar to Iranian Jews and Levantine non-Jews. Their Indian mtDNA haplogroups M5a1, M50, R5a2b, R6a2, and R8 are absent from other Jewish populations and are also dramatically different from the haplogroups H13a2a and M39a1 that are found among the Bene Israel Jews of India.[712] However, the mtDNA haplogroups M30c1a1 and R30b, both also of Indian origin, are shared by Cochin Jews and Bene Israel Jews. Cochin Jews share the non-Indian haplogroup K1a1b1a with Ashkenazic Jews.

The Krymchaks from the Crimean peninsula are Rabbinical Jews who descend from both Western Jews (Greek-speaking Jews, Turkish Jews, Spanish Jews, Italian Jews, and Ashkenazic Jews) and Mizrahi Jews (Persian Jews and Jews from the Caucasus).[713] Their autosomal DNA reveals a mixture of predominantly Middle Eastern and European elements with small amounts of North African and East Eurasian elements. Leo Cooper informed me that some Krymchaks who tested with 23andMe belong to the mtDNA haplogroups H2, T1, and U5b2a1,[714] the last one reminding us of the full Ashkenazic sequence U5b2a1a. As of the time of this writing, I do not have any complete mitochondrial sequences for Krymchaks who may have tested at Family Tree DNA and there are no published studies on them.

Karaite Jews and Rabbinical Jews had a religious schism that was caused in large part by the Karaites' rejection of the legal authority of the Talmud and Mishnah. As a result, Karaite Jews seldom intermarried with Rabbinical Jews in the thirteenth-nineteenth centuries.[715] Modern Karaite Jewish communities have included Iraqi Karaite, Egyptian Karaite, Crimean Karaite, Galician Karaite,[716] and Lithuanian Karaite.[717] Karaite Jews from Eastern Europe largely descend from Karaite Jews from Turkey who moved there during the Byzantine and Ottoman eras.[718] In my study of the genetics of European Karaites, which showed that many of their Y-DNA lineages are Middle Eastern and related to those of other kinds of Jews (including Ashkenazim),[719] I discovered a mtDNA haplogroup, H9a, in a Crimean Karaite that Ashkenazic Jews never carry. Judging by its closest maternal matches, Crimean Karaites apparently inherited H9a from a European non-Jewish ancestor.[720] Deeper analysis of a Crimean Karaite sample from my study who was partially sequenced into mtDNA haplogroup H by Family Tree DNA[721] and later into H2a1[722] by 23andMe led Ian Logan and me to determine that he belongs specifically to haplogroup H2a1i, which has his HVR1 mutation G16145A and his HVR2 mutation C113T.[723] Other H2a1i carriers are in Turkey,[724] Iran,[725] Bulgaria, Greece, and Italy[726] but Ashkenazim do not carry it.

Chueta lineages that are not shared by Ashkenazic Jews

Also called Xuetas, the Chuetas of the Spanish island of Mallorca (Majorca) are descendants of Sephardic Conversos (Jews who converted, or outwardly pretended to convert, to Roman Catholicism) who stayed largely endogamous over the centuries. While R0a2m, K1a1b1a, and HV0 are part of the gene pools of both Chuetas and Ashkenazim, other maternal haplogroups, including but not limited to J1c2o, J1d1, J2a1a1, K1c, K2b1a1, L3eb+152, M5a1, T2b5a1, T2b23, T2c1d, U1a1a, U5b1d2, and U5b3, are found in Chuetas[727] but never in Ashkenazim. The presence of M5a1 in both Chuetas and Cochin Jews from India is probably a coincidence considering its South Asian (not Middle Eastern) origin and that it is also found in Romany people who settled in Europe (including Spain).

Ferragut's genetics team observed that the Chuetas' haplogroups R0a2m, K1a1b1a, U1a1a, and L3eb+152 are "rarely found in neighbouring populations" and that this "could also mean that they might have been present in the original Jewish Majorcan gene pool."[728]

Bragança Crypto-Jewish lineages that are not shared by Ashkenazic Jews

There are many people living in northeastern Portugal's town of Bragança and the district of the same name who descend from Jews. It is known that Jews already lived in the town by 1279. All Jews in Portugal were forced to convert to Roman Catholicism in 1497 but their descendants in Bragança managed to retain some Jewish customs and beliefs and continued to practice endogamous marriage. While N1b1a2, T2e1b, and U2e1a1 are shared by Bragança Crypto-Jews and by Ashkenazic Jews, the Bragança

Crypto-Jews possess some maternal haplogroups that Ashkenazim never carry, including HV0b, N1a1a1a2, N1b1a5, and T2b11. Scientists believe that the Bragança Crypto-Jews inherited all seven of these lineages from Sephardic forebears and did not find identical mitochondrial sequences in the general Portuguese population.[729]

Belmonte Crypto-Jewish lineages that are not shared by Ashkenazic Jews

A whopping 93 percent of the Crypto-Jews living in the small municipality of Belmonte in northeastern Portugal belong to the mtDNA haplogroup HV0b, which as stated earlier is also found among the Bragança Crypto-Jews[730] but not among Ashkenazic Jews. The only other mtDNA haplogroup that scientists know that the Belmonte Crypto-Jews possess is H[731] but so far their specific subclade of H has not yet been reported.

Polish lineages that are not shared by Ashkenazic Jews

Although multiple Ashkenazic maternal haplogroups (H3p, K1a1b1a, K1a9, K2a2a1, and L2a1l2a) exist among ethnic Poles as the result of Jewish conversions to Christianity, and several Polish maternal haplogroups of European origin (including H11a2a2, H11b1, and W3a1a1) exist among Ashkenazim as the result of conversions of Slavic (mostly Polish and possibly Czech) women to Judaism, there are many other Polish haplogroups that Ashkenazim never carry. These include C5c1, D5a2a1a1, H5a1f, H20a3, H35, H49, H94, I3, L0a1a, L3e, U2e1b1, W1g, W5, and Z1a1a, among many others.

German lineages that are not shared by Ashkenazic Jews

H2a1e1a, H4a1a1a, H4a1a3a, H5a1, H6a1b2, H6a1b3, H7e, H7j, H13a1a1a, H26c, H56, H65a, J1b1a1, J1c4, J1c7a, T1a1, T2b25, U3a1a, U4a3a, U5b2a1a, and V1a1 are among the haplogroups that ethnic Germans from Germany share with Ashkenazim, sometimes because German women had converted to Judaism. However, there are many other haplogroups that only non-Jewish Germans have, such as H1bv1, H5a2, H10b, H16b, H24a, I2f, and T2b5.

Italian lineages that are not shared by Ashkenazic Jews

Autosomal DNA evidence suggests that there was genetic input from Italians into the early Western Jewish ancestors of Ashkenazic Jews more than a thousand years ago. As stated earlier, J2-Y33795 may be an example of an Ashkenazic paternal haplogroup of Italian origin. There are, in addition, multiple mtDNA haplogroup subclades that are shared by modern mainland Italian Christians as well as by Ashkenazic Jews, including H1bo, H3ap, H5a7, H7c2, J1c3e2, J1c13, K1a4a, U8b1b1, and possibly H1b1a, that might have Italian origins for both populations.

Haplogroups H13a1a1, H47, HV0-T195C!, HV1b2, J1b1a1, J1c-C16261T, J1c1, K1a4a, K2a*, T2b25, and T2g1a are shared by modern Sicilians and Ashkenazim. Not all of these shared haplogroups necessarily have common non-Jewish ancestors from Italy because many Sicilians from the Palermo region in the northwest, the Siracusa region in the southeast, and the province of Agrigento in the southwest and some Italians from the farthest south regions of the mainland descend in part from Sicilian Jews and Sephardic Jews who were forced to convert to Roman Catholicism in the 1490s. There had also been ancient Phoenician settlers in Sicily who were genetically similar to the Israelites.

On the other hand, there are many other modern Italian haplogroups that Ashkenazim never carry, such as H1c4b, H20a2, K1b2a3, U1a3, U2e2a, and X2d1. Sicilian haplogroups that Ashkenazim never carry include HV2, I6a, J1d5, K1c1f, L3e2, T1a11, U7a3a, and others. Sardinians have many haplogroups that Ashkenazim lack, such as K1a2d, U1a1c1, U1a1c3, U5b1i1, and U6d1a.

Greek lineages that are not shared by Ashkenazic Jews

A portion of the southern European ancestry in Ashkenazic Jews appears to derive from Greeks. A Greek ancestor appears to be the source for at least one Ashkenazic Y-chromosomal lineage (I-Y23115). There are some clues from autosomal DNA that Greeks could have had a sizeable impact, but input from Greek women is not certain. H1u2, H41a, K1a4a, J1c-C16261T, J1c13, T2a1b, T2b4, V1a1, and X2e2a are maternal haplogroups that are shared by modern non-Jews from Greece and Ashkenazic Jews, but none of these is exclusive to the two groups. Some of the many modern Greek haplogroups that Ashkenazim never carry include H14a2, H15a1b, H55b, H101, HV4b, T2a1a, and W6c1a. Modern Greeks have preserved much of the genetics of the ancient Greeks but there are some differences in mainland Greeks resulting from medieval inputs into the Greek population from other Balkan peoples. Greek islanders have remained more genetically similar to the ancient Greeks.

Lebanese lineages that are not shared by Ashkenazic Jews

Ashkenazim and Lebanese people are both descended from the Canaanite people of the ancient Levant, with Lebanese people (especially Lebanese Christians) having preserved almost completely Levantine genetic profiles up through the present time. The Lebanese descend from the branch of Canaanites who are known as the Phoenicians. In part due to the high number of new maternal lineages that Ashkenazim accumulated in the Jewish diaspora, there are many mtDNA haplogroups in Lebanon that Ashkenazim never have, including but not limited to H2a2a1, H13b, H20a1, H33b, J1b4a1, J1c4b, R0a2c, U1b3, and the Druze lineage X2f.

Samaritan lineages that are not shared by Ashkenazic Jews

The Samaritan people are genetically an entirely Middle Eastern people of Israel who have preserved a large proportion of Israelite ancestry and have also preserved a version of the Paleo-Hebrew alphabet and a partly different version of the Torah. Over the centuries, the Samaritan population dramatically decreased in numbers and lost much of their genetic diversity (including haplogroup diversity) due in part to conversions of Samaritans to Islam and, more recently, to Judaism.

As it happens, Ashkenazim never carry the authentically old Samaritan mtDNA haplogroups T2a1b1a, U7a1, and U7b. Other Middle Eastern peoples with varieties of U7b include Israeli Druze, Kuwaitis, and Assyrians.

To help to grow their population once again and to decrease the prevalence of Samaritan genetic diseases, Samaritan men have recently been marrying non-Samaritan women who converted to Samaritanism, including former Ashkenazic Jews and Ukrainian Christians.

Lineages from ancient Israel that are not shared by Ashkenazic Jews

Likewise, there are some ancient samples from Israel that carried mtDNA haplogroups that Ashkenazim did not inherit. While T2g1a and HV1a'b'c were found in Copper Age inhabitants of Israel and are also found in Ashkenazim to this day, the following Copper Age haplogroups from Israel are not shared by Ashkenazim: J2a2d, N1a1b, and T1a2. The Bronze Age haplogroup H40a and the Neolithic-era haplogroup K1a4b from Israel are also absent from the Ashkenazic gene pool.

Chapter 4: Conclusion

The uniparental DNA evidence presented in this book is consistent with the autosomal DNA studies that show that Ashkenazim are indigenous to both Europe and the Middle East in major ways, and to North Africa and China in much smaller ways.

The initial Ashkenazic population of early medieval Germany would have been founded by women who were already Jewish and whose maternal haplogroups included K1a1b1a, K1a4a, K1a9, K2a2a1, L2a1l2a, M1a1b1c, N1b1b1, R0a4, U1b1, and U7a5, among others. Many of these haplogroups are associated with the Middle East and Mediterranean regions. These haplogroups are found in German Jews who were born in the twentieth century but have deeper roots in Germany. Originally, I thought that it would probably never be possible to absolutely confirm that medieval German Jews had these haplogroups, because Jewish rabbinical law (*halakha*) forbids disturbing or digging up the graves of Jews and, as a result, archaeologists and geneticists working in Europe sometimes respect this prohibition and do not study identifiably Jewish remains in genetic laboratories. For the same reason, I had thought it would probably not be possible to ascertain whether any haplogroups that early German Jews possessed later died out in the Ashkenazic population. However, it came to my attention that the haplogroups of some late-medieval Jews from a certain German city have in fact been genetically sequenced, but I am not allowed to mention any specifics before they are published.

The mitochondrial haplogroup H7j presents clear-cut evidence of a West Germanic ancestor for Ashkenazim, probably North German in particular, which is surprising. Other haplogroups that would have come from German converts to Judaism include H2a2b1, H4a1a1a, J1c7a, and probably H5c2, H6a1a3a, H6a1b3, H7e, H26c, J1c4, and T2b16. H1f, H1ai1, and T1a1k1 also came from northern European converts. H13a1a1a and T2b25 could come from either German or Slavic women. Slavic, probably Polish, women transmitted multiple haplogroups, including H11a2a2, H11b1, V7a, and W3a1a1 and possibly H1b, H5a1, H11a1, H40b, H41a, U3a1a, U5a1d2b, U5b1e1, V1a1, and V7b to Ashkenazim. U5a2b2a and W3b1 came from central European converts. Haplogroup H2a1e1a could come from either a German or French woman, more likely the former.

The foregoing mitochondrial DNA evidence confirms that Ashkenazic Jews have partial German and Polish origins. Ashkenazic culture did not internalize these identities due to Judaism's emphasis on the more significant portion of Ashkenazic ancestry that derives from ancient Israel and the notion that a convert is reborn into the people of Israel and given equal status to Israelite Jews in most respects. Knowledge of a convert in the family was not passed along through many generations. Most Polish Jews understandably regarded themselves as a distinct people compared to their Polish Catholic neighbors, except for a small minority of Polish Jews who became fluent in the Polish language and participated in mainstream Polish culture.[732] Anti-Jewish ideologies and actions by the Catholic Church in Poland broadened the divide between Ashkenazim and Poles and prevented them from developing a unified identity.[733] Many German Jews who belonged to the Reform denomination and were deeply integrated into German society during the nineteenth and early twentieth centuries regarded themselves as Germans of the Jewish faith, a view that Orthodox Jews did not agree with, and did not feel close to Eastern European Jews due to cultural and religious differences.[734] Some German-speaking Jews who lived in Galicia and Bukovina likewise considered themselves to be Germans,[735] but Yiddish-speaking Jews in Galicia usually did not. At the same time, according to Kateřina Čapková, some Ashkenazim in Bohemia who did not practice Judaism but spoke the Czech language "and were integrated into the Czech cultural and social environment" thought of themselves as "Czechs of Jewish descent,"[736] which did acknowledge their Israelite origins.

Haplogroups H1bw and X2b7 possibly came to Ashkenazim from French women but current evidence shows closer Spanish ties for the Ashkenazic variety of X2b7.

A French person who converted to Judaism could have been responsible for bringing Tay-Sachs disease's defective allele into the Ashkenazic community, since Tay-Sachs is also found among Eastern Quebecois and Acadians in Canada and among Cajuns in Louisiana. Whereas Acadians and Quebecois in the late twentieth century carried a different Tay-Sachs mutation than the Ashkenazic version, Cajuns and Ashkenazim share the same mutation that causes the infantile form of the disease.[737] The Cajun-Ashkenazic mutation is referred to as the beta-hexosaminidase A alpha-subunit exon 11 insertion and is also called 1278insTATC. Most of the Cajun carriers of the mutation genealogically trace back to a particular non-Jewish ancestral couple who lived in eighteenth-century France. 1278insTATC is one of the five mutations causing Tay-Sachs that have been found among Iraqi Jews,[738] but this is presumably because carriers of that mutation had an Ashkenazic ancestor. Genetic analysis has established that modern French Catholics do not descend from Ashkenazic, Sephardic, Mizrahi, or Italki Jews, and this is the reason that the direction of the introgression of Tay-Sachs (into Jews rather than from Jews) is clear. The particular mutation that causes this disease in Ashkenazim probably originated in France. This tentatively suggests that at least one French woman may have contributed mtDNA to Ashkenazim, but more research is needed.

Surprisingly, most of the Ashkenazic mitochondrial DNA lineages of European origin do not have particular connections to Italy, contradicting others' speculations that all or most of these lineages were supplied by central or northern Italian women who had converted to Judaism.[739] However, Ashkenazic haplogroups like H1bo, H7c2, I5a1b, J1c3e2, J1c13, K1a4a, T2a1b, and U8b1b1 might have originated with proselyte women in ancient Italy, although J1c13, K1a4a, and T2a1b might have come from Greek women instead. It is currently uncertain whether or not the haplogroups H47, HV1a'b'c, J2b1e, N1b1a2, T2g1a, and U1b1 originated from Italian converts; it is more likely that Israelite women passed them along to the Ashkenazim. J1c-C16261T could come from an Italian, a Greek, or an Israelite.

Other mitochondrial DNA lineages (including HV1b2, U7a5, R0a2m, H6a1a1a, N1b2, K2a2a1, and K1a9) present clear evidence of the partial Middle Eastern, including Judean, origins of the Ashkenazic Jews and therefore serve as the maternal complements to the many Ashkenazic

Y-chromosomal DNA lineages with deep roots in Israel and the rest of the Middle East. All of those mitochondrial haplogroups probably came to Ashkenazim from ancient Israel although HV1b2 could have originated with a convert to Judaism in Mesopotamia. This shows that some Jewish women of Judean origin had moved to Europe, disproving the claim that all Ashkenazic matrilines come from European converts to Judaism.

The Ashkenazic mitochondrial DNA haplogroup U6a7a1b derived from a Berber (Amazigh) ancestor in North Africa who converted to Judaism, and L2a1l2a likely did also. The women who brought these haplogroups in combination with a few male Berber counterparts caused modern Ashkenazim to typically score 2 to 4 percent North African in autosomal DNA calculators like Eurogenes K36. Much of this ancestry probably originated with the Jews who lived in the Cyrenaica region in what is now eastern Libya since some Jews from there later moved to Rome.

Ashkenazim inherited the mitochondrial DNA haplogroup M33c from a Chinese woman, probably one from the Han ethnicity. I previously thought that their haplogroups N9a3 and A12'23 also came from Chinese women but we need to reopen the possibility that one or both of those could have been Khazarian because of the close Bashkir, Chechen, and Ingush matches to the former and the close Uzbekistani Turkmen and ancient Central Asian matches to the latter. All three of these haplogroups were certainly from East Asia originally. The women carrying these haplogroups caused the characteristically East Asian 1540C allele of the ectodysplasin A receptor (EDAR) gene to exist among some modern Eastern Ashkenazim, with an overall prevalence of this allele among 1.7 percent of Ashkenazim. 1540C is especially common among modern East Asians, including the Japanese and Qiang peoples. One of the effects of this allele is to increase the thickness of a carrier's scalp hair.[740] Most East European Jews have 1 or 2 percent of East Asian autosomal DNA admixture.

No Ashkenazic mtDNA or Y-DNA lines have been confirmed to come from the ethnic Khazars, a branch of the Turkic peoples who founded the kaganate of Khazaria that existed in southern Russia and eastern Ukraine from 652 until 969. Many ethnic and religious groups lived in Khazaria.[741] Some of Khazaria's Jewish residents were descended from the Israelites while others were recent converts to Judaism who belonged to the Khazar and Alan tribes.[742] Skeletons from medieval Khazaria have been studied by both archaeologists and geneticists and many of these have been ethnic Khazars.[743] The ethnic Khazarian mtDNA haplogroups identified by Tatiana Tatarinova's genetics team were C4, C4a1, C4a1c, D4b1a1a, D4e5, H1a3, H5b, H13c1, and X2e and in their 2019 preprint "Diverse genetic origins of medieval steppe nomad conquerors" they wrote that none of the specific subclades or haplotypes are found in Ashkenazim.[744] Although X2e2a is carried by some Ashkenazim, it apparently does not exactly match the Khazarian variety. A 2015 study by Afanasyev's team identified I4a and D4m2 in members of Khazaria's Saltovo-Mayaki culture.[745] Neither of those is an Ashkenazic haplogroup. Khazarian A12'23, H40b, N9a3, and U5a1d2b samples have not been found to date. Certain Hungarian Jewish scholars, notably including Rabbi Sámuel Kohn (1841–1920), argued that Ashkenazic Jews in Hungary descend from the Khazars and the Magyars.[746] Other Jewish scholars extended the Khazar origin hypothesis to other Eastern Ashkenazic subgroups.

The Ashkenazic Litvak Y-DNA haplogroup G2a-FGC1093 is shared with North Ossetians, who are descendants of the Alans, and this shows that some Ashkenazim descend to a minor degree from the Alans. There is not similar evidence when it comes to mtDNA. The 2013 study by Behar's team that genetically compared Ashkenazim to many other populations, such as North Ossetians, Chechens, and some Turkic peoples, was not able to detect any large identical-by-descent autosomal DNA segments shared by Ashkenazim and those groups because, as they acknowledged, their identical-by-descent detections using the minimum segment length that they chose only caught connections within the last 20 generations or so,[747] post-dating by several centuries the time when certain Alans and Khazars became Jews and formed separate subpopulations. Accordingly, they wrote that they "cannot rule out the possibility that a level of Khazar or other Caucasus admixture occurred below the level of detectability in our study."[748] If there is any Khazar ancestry in Ashkenazim at all, it would be represented by Leo Cooper's autosomal DNA models that predict that Ashkenazim from Poland descend 0.6 percent from Iron Age nomads from Kazakhstan and that Ashkenazim from Ukraine descend 0.2 percent from them, but it is possible that these are noise readings, and his runs found that Ashkenazim from Lithuania and

Belarus score 0 percent in that element.[749] On the other hand, his models confirm that all Eastern Ashkenazim definitely have Han-related DNA from southern China.

Some medieval Alans carried the East Eurasian mtDNA haplogroups C4a, D4, and F2c.[750] None of those are found in Ashkenazim.

None of the Ashkenazic mtDNA haplogroups derived from a South Asian woman who converted to Judaism.

Based on their limited geographic distributions among Ashkenazic subgroups and their cross-ethnic affinities, Cooper has plausibly suggested that mtDNA haplogroups like N9a3 and M33c along with Y-DNA haplogroups like G2a-FGC1093, I-BY424, J-FGC75679, and R2a-FGC13211[751] among modern Eastern Ashkenazim may be remnants from the East Slavic-speaking pre-Ashkenazic Jewish population of eastern Europe, the East Knaanim. That being said, all Ashkenazim also descend from the ninth-century founders of the Ashkenazic population in Germany, just not only from them. With regard to Belarus in particular, Alexander Beider wrote, "Few in number, Slavic-speaking Jews merged with the newly arriving Yiddish-speaking migrants from the West during the sixteenth and the first half of the seventeenth centuries."[752] Their intermarriage explains why Belarusian Jews came to carry many of the same paternal and maternal haplogroups as German Jews.

It is possible that Ashkenazim picked up their Sephardic-associated haplogroups H25, I5a1b, N1b1a2, U2e1a1, and U6a7a1b from Sephardic Jewish women with roots in Spain who moved from Turkey to Ukraine. Although H1bo, H6a1a1a, H56, R0a2m, R0a4, T1a1j, T1b3, and T2e1b and the latter's subclade T2e1b1 also have Sephardic associations, it is less likely that Ashkenazim inherited those from Sephardim. Perhaps Ashkenazim inherited some or all of these from the early Jews in southern Europe who were simultaneously proto-Sephardic and proto-Ashkenazic. That was surely the case for the very widespread haplogroups K1a1b1a and N1b1b1.

Genetic testing has brought to light several aspects of Ashkenazic origins that would not have otherwise been known had we relied solely on findings from documentation, religious tradition, linguistics, onomastics, gravestones, or archaeology. Population genetics is a tool that historians can use to fill in the gaps.

Notes

1   Leonard B. Glick, *Abraham's Heirs: Jews and Christians in Medieval Europe* (Syracuse, NY: Syracuse University Press, 1999), 29.

2   Jits van Straten, *Ashkenazic Jews and the Biblical Israelites: The Early Demographic Development of East European Ashkenazis* (Berlin: De Gruyter Oldenbourg, 2021), 1-2.

3   Glick, *Abraham's Heirs*, 44.

4   Alexander Beider, *Origins of Yiddish Dialects* (Oxford: Oxford University Press, 2015), 426, 533, 562, 567; Robert Dittmann, "West Slavic Canaanite Glosses in Medieval Hebrew Manuscripts," *Judaica: Beiträge zum Verstehen des Judentums* 73, no. 2-3 (June-September 2017): 243, 282.

5   Kevin Alan Brook, *The Jews of Khazaria*, 3rd ed. (Lanham, MD: Rowman & Littlefield, 2018), 196. They were probably not descended from the 3rd-century Jews of Roman-ruled Pannonia, which included western Hungary (see ibid., 77-78).

6   Ibid., 192-193; Alexander Kulik and Judith Kalik, "The Beginnings of Polish Jewry: Reevaluating the Evidence for the Eleventh to Fourteenth Centuries," *Zeitschrift für Ostmitteleuropa-Forschung* 70, no. 2 (2021): 157, 163, 165-166.

7   Brook, *The Jews of Khazaria*, 95, 190.

8   Ibid., 173, 175.

9   Beider, *Origins of Yiddish Dialects*, 427, 435.

10   Alexander Beider, "Names and Naming," in *The YIVO Encyclopedia of Jews in Eastern Europe*, ed. Gershon D. Hundert (New Haven, CT: Yale University Press, 2008), vol. 1, 1248-1251.

11   Alexander Avraham, "Sephardim," in *The YIVO Encyclopedia of Jews in Eastern Europe*, ed. Gershon D. Hundert (New Haven, CT: Yale University Press, 2008), vol. 2, 1689-1692; Wacław Wierzbieniec, "Zamość," in *The YIVO Encyclopedia of Jews in Eastern Europe*, ed. Gershon D. Hundert (New Haven, CT: Yale University Press, 2008), vol. 2, 2111-2112; Alexander Beider, "Exceptional Ashkenazic Surnames of Sephardic Origin," *Avotaynu: The International Review of Jewish Genealogy* 33, no. 4 (Winter 2017): 3-5.

12   Kevin Alan Brook, "Sephardic Jews in Belarus," *ZichronNote: The Journal of the San Francisco Bay Area Jewish Genealogical Society* 38, no. 1-2 (February/May 2018): 5-6; Kevin Alan Brook, "Sephardic Jews in Lithuania and Latvia," *ZichronNote: The Journal of the San Francisco Bay Area Jewish Genealogical Society* 36, no. 3 (August 2016): 9-11. As a result, triangulating autosomal DNA segments of Sephardic origin with lengths between 6 and 14 centimorgans are ubiquitous in the contemporary

Eastern Ashkenazic population and simultaneously shared by members of diverse populations such as Mexican Catholics, Peruvian Catholics, Moroccan Jews, and Ashkenazic Jews within the past 16 generations. I spent years finding and validating hundreds of these segments. Far fewer Ashkenazim have uniparental lineages of Sephardic origin but an example of one is Bennett Greenspan's Y-DNA line, which matches non-Jewish Hispanics, including a New Mexican Hispano, in his 67-marker and 111-marker STR screens at the company he founded, Family Tree DNA.

13   Edna Engel, "Immigrant Scribes' Handwriting in Northern Italy from the Late Thirteenth to the Mid-Sixteenth Century: Sephardi and Ashkenazi Attitudes toward the Italian Script," in *The Late Medieval Hebrew Book in the Western Mediterranean: Hebrew Manuscripts and Incunabula in Context*, ed. Javier del Barco (Leiden: Brill, 2015), 28.

14   K. E. Fleming, *Greece—A Jewish History* (Princeton, NJ: Princeton University Press, 2008), 9.

15   In post #10690 in Anthrogenica thread https://anthrogenica.com/showthread. php?14484-Could-Western-Jews-(Ash-and-Seph-)-descend-from-Aegeans-and-Levantine-admixture/page1069 dated March 10, 2021, accessed January 21, 2022, Leo R. Cooper presented the results of a model he worked on for Ashkenazim from Germany that showed them to be 42.4 percent like Middle/Late Bronze Age samples from Tel Megiddo.

16   On the autosomal similarity of Ashkenazim to Greeks from Crete, see: Petros Drineas, Fotis Tsetsos, et al., "Genetic history of the population of Crete," *Annals of Human Genetics* (June 13, 2019).

17   James Xue, Todd Lencz, et al., "The Time and Place of European Admixture in Ashkenazic Jewish History," *PLoS Genetics* 13, no. 4 (April 4, 2017).

18   The particular Romaniotes from Ioannina who were used for these models by Ariel Lomes had ancestors who did not intermarry with Sephardim, as is known from their documented genealogies back to the early to mid- 1800s, but other Romaniotes did intermarry with Sephardim, especially after World War II.

19   Ariel Lomes, post #3730 dated October 13, 2018, accessed January 21, 2022, in Anthrogenica, https://anthrogenica.com/showthread.php?14484-Could-Western-Jews-(Ash-and-Seph-)-descend-from-Aegeans-and-Levantine-admixture/page373

20   Siiri Rootsi, Doron M. Behar, et al., "Phylogenetic Applications of Whole Y-Chromosome Sequences and the Near Eastern Origin of Ashkenazi Levites," *Nature Communications* 4 (December 17, 2013).

21   Doron M. Behar, Lauri Saag, et al., "The genetic variation in the R1a clade among the Ashkenazi Levites' Y chromosome," *Scientific Reports* 7 (November 2, 2017): "Results" section.

22   The expandable phylogenetic tree at https://www.yfull.com/tree/ is the source for this and all subsequent references to "YTree," YFull, accessed 2021-2022. All of YFull's "ybp" (years before present) estimates designate 1950 as the "present" year.

23   Michael F. Hammer, Doron M. Behar, et al., "Extended Y chromosome haplotypes resolve multiple and unique lineages of the Jewish priesthood," *Human Genetics* 126, no. 5 (November 2009): 707-717.

24   Vladimir Gurianov, Dmitry Adamov, et al., "Clarification of Y-DNA Haplogroup Q1b Phylogenetic Structure Based on Y-Chromosome Full Sequencing," *Russian Journal of Genetic Genealogy* 7, no. 1 (March 31, 2015): 94.

25 Brook, *The Jews of Khazaria*, 235. However, this is the only surviving medieval document that mentioned Alans becoming Jewish.

26 Murat Dzhaubermezov, Liliia Gabidullina, et al., "Ancient DNA analysis of Early Medieval Alan populations of the North Caucasus" (paper presented at Widening Horizons: 27th Annual Meeting of the European Association of Archaeologists, Kiel, Germany, September 8-11, 2021), Figure 4.

27 Leo R. Cooper, post #66 dated September 14, 2021, accessed September 16, 2021, in Anthrogenica, https://anthrogenica.com/showthread.php?23847-How-much-Middle-Eastern-ancestry-do-Ashkenazim-have/page7

28 Margaret L. Antonio, Ziyue Gao, et al., "Ancient Rome: A genetic crossroads of Europe and the Mediterranean," *Science* 366, no. 6466 (November 8, 2019): Table S2 where he identified as sample R474. Although his haplogroup is listed there as "J-M12," it was later fully sequenced as J2-Y45181 by other researchers, including those at YFull.

29 For example, see the genetic analysis in Rachel Unkefer, "Disproving a Cossack Paternal Ancestry for an Ashkenazic Lineage," *Avotaynu* 36, no. 3 (Fall 2020): 47-48 that chronologically supports a Greek origin but disproves a Cossack origin for Y-DNA haplogroup I-P37.

30 David Wesolowski, "Estonian Biocentre Human Genome Diversity Panel (EGDP)," September 26, 2016, http://eurogenes.blogspot.com/2016/09/estonian-biocentre-human-genome.html

31 Pau Figueras, "Epigraphic Evidence for Proselytism in Ancient Judaism," *Immanuel* 24/25 (1990): 199.

32 Ibid., 199.

33 Ibid., 200.

34 Ibid., 200.

35 Rami Reiner, "Tough are Gerim: Conversion to Judaism in Medieval Europe," *Havruta: A Journal of Jewish Conversation* 1 (Spring 2008): 56.

36 Robert Chazan, *The Jews of Medieval Western Christendom: 1000-1500* (Cambridge: Cambridge University Press, 2006), 229.

37 Wayne Allen, *The Cantor: From the Mishnah to Modernity* (Eugene, OR: Wipf and Stock, 2019), 138.

38 Herman Rosenthal, "Lithuania," in *The Jewish Encyclopedia*, ed. Isidore Singer (New York: Funk & Wagnalls, 1904), vol. 8, 126.

39 Brook, *The Jews of Khazaria*, 238-239; Gershon D. Hundert, *Jews in Poland-Lithuania in the Eighteenth Century: A Genealogy of Modernity* (Berkeley, CA: University of California Press, 2004), 51; Jacob Goldberg, "Tavernkeeping," in *The YIVO Encyclopedia of Jews in Eastern Europe*, ed. Gershon D. Hundert (New Haven, CT: Yale University Press, 2008), vol. 2.

40 Magda Teter, *Jews and Heretics in Catholic Poland: A Beleaguered Church in the Post-Reformation Era* (Cambridge: Cambridge University Press, 2006), 67.

41 Doron M. Behar, Michael F. Hammer, et al., "MtDNA Evidence for a Genetic Bottleneck in the Early History of the Ashkenazi Jewish Population," *European Journal of Human Genetics* 12, no. 5 (May 2004): 355-364.

42 Ellen Coffman-Levy, "A mosaic of people: the Jewish story and a reassessment of the DNA evidence," *Journal of Genetic Genealogy* 1 (2005): 12-33.

43 Marta D. Costa, Joana B. Pereira, et al., "A Substantial Prehistoric European Ancestry amongst Ashkenazi Maternal Lineages," *Nature Communications* 4 (October 8, 2013).

44 Genetic distance in this context is always based on the coding region for a complete mtDNA sequence and refers to "the number of differences, or mutations, between two sets of results. A genetic distance of zero means there are no differences in the results being compared against one another, i.e., an exact match." ("Genetic Distance," Family Tree DNA, accessed January 22, 2022, https://learn.familytreedna.com/faq-items/genetic-distance/)

45 Wim Penninx, "Catalogue of mtdna Jewish branches," 2019, accessed October 3, 2021, https://jewishdna.net/Mtdna.html

46 Behar, Hammer, et al., "MtDNA Evidence for a Genetic Bottleneck...," "Supplementary Data."

47 The expandable phylogenetic tree at https://www.yfull.com/mtree/ is the source for this and all subsequent references to "MTree," YFull, accessed 2021-2022. For example, the section for haplogroup A-a1b is https://www.yfull.com/mtree/A-a1b/

48 The expandable phylogenetic tree at https://www.familytreedna.com/public/mt-dna-haplotree/ is the source for this and all subsequent references to "mtDNA Haplotree," Family Tree DNA, accessed 2021-2022.

49 Min-Sheng Peng, Weifang Xu, et al., "Mitochondrial genomes uncover the maternal history of the Pamir populations," *European Journal of Human Genetics* 26, no. 1 (January 2018): supplementary data. These samples' GenBank codes are MF522991 and MF523016.

50 Wibhu Kutanan, Jatupol Kampuansai, et al., "Complete mitochondrial genomes of Thai and Lao populations indicate an ancient origin of Austroasiatic groups and demic diffusion in the spread of Tai-Kadai languages," *Human Genetics* 136, no. 1 (January 2017): supplementary data. This sample's GenBank code is KX456634.

51 Kristiina Tambets, Bayazit Yunusbayev, et al., "Genes reveal traces of common recent demographic history for most of the Uralic-speaking populations," *Genome Biology* 19 (September 21, 2018): Table S3 where they are listed with the haplogroup "A12/23".

52 Geoffrey Sea, "Haplogroup discovery," accessed March 4, 2022, https://geoffreysea.com/index.php/pages/about-us; Geoffrey Sea, "Haplogroup A," Facebook, February 6, 2022, accessed April 13, 2022, https://www.facebook.com/groups/AdenaCore/permalink/4925500514177395/

53 Miroslava Derenko, Galina Denisova, et al., "Mitogenomic diversity and differentiation of the Buryats," *Journal of Human Genetics* 63, no. 1 (January 2018): supplementary data. This sample is coded as MF043437 in GenBank.

54 Miroslava Derenko, Boris A. Malyarchuk, et al., "Phylogeographic analysis of mitochondrial DNA in northern Asian populations," *American Journal of Human Genetics* 81, no. 5 (November 2007). These samples are coded as EF153791 and EF153784 in GenBank.

55 Dan Zhao, Yingying Ding, et al., "Mitochondrial Haplogroups N9 and G Are Associated with Metabolic Syndrome Among Human Immunodeficiency Virus-Infected Patients in China," *AIDS Research and Human Retroviruses* 35, no. 6 (June 2019). Their GenBank codes are MH553660, MH553811, and MH553825.

56 Costa, Pereira, et al., "A Substantial Prehistoric European Ancestry...," Figure 7, which only includes non-Ashkenazic carriers of this haplogroup.

57 Personal communication with Leo Cooper, May 6, 2021.

58 Ashot Margaryan, Daniel J. Lawson, et al., "Population genomics of the Viking world," *Nature* 585, no. 7825 (September 16, 2020). He is identified as sample VK202 in this study.

59 "Jewish Ukraine West - mtDNA Test Results for Members," Family Tree DNA, accessed January 13, 2022, https://www.familytreedna.com/public/Jewish_Ukraine_West?iframe=mtresults

60 Public family tree of Wendy L. Wood Callahan, WikiTree, accessed September 11, 2021, https://www.wikitree.com/wiki/Wood-22154

61 Personal communication with Leo Cooper, March 27, 2021.

62 "Jewish Ukraine West - mtDNA Test Results for Members."

63 Sanni Översti, Kerttu Majander, et al., "Human mitochondrial DNA lineages in Iron-Age Fennoscandia suggest incipient admixture and eastern introduction of farming-related maternal ancestry," *Scientific Reports* 9 (November 15, 2019): supplementary data. The relevant sample is coded as MN540564 in GenBank.

64 Costa, Pereira, et al., "A Substantial Prehistoric European Ancestry...," Figure 7, which only includes non-Ashkenazic carriers of this haplogroup.

65 This sample is coded as KC286588 in GenBank.

66 Martin Bodner, Alessandra Iuvaro, et al., "Helena, the hidden beauty: Resolving the most common West Eurasian mtDNA control region haplotype by massively parallel sequencing an Italian population sample," *Forensic Science International: Genetics* 15 (March 2015). The relevant sample is coded as KM252737 in GenBank.

67 Nick Patterson, Michael Isakov, et al., "Large-scale migration into Britain during the Middle to Late Bronze Age." *Nature* 601, no. 7894 (January 27, 2022): Supplementary Table 1 listing her sample code I19874.

68 Jewish vital records from Poland and western Ukraine often allow tracing back to the 1780s-1820s.

69 Personal communication with Leo Cooper, May 29, 2021.

70 This sample is coded as MT588290 in GenBank.

71 Penninx, "Catalogue of mtdna Jewish branches."

72 Relevant samples are coded as MK059553 (alternative identification code LP373), MK059707 (alternative identification code LM45), and MG429047 respectively in GenBank.

73 Alessandro Achilli, Chiara Rengo, et al., "The Molecular Dissection of mtDNA Haplogroup H Confirms That the Franco-Cantabrian Glacial Refuge Was a Major Source for the European Gene Pool," *American Journal of Human Genetics* 75, no. 5 (December 2004): supplemental data. This Italian sample is coded as AY738975 in GenBank.

74 Boris A. Malyarchuk, Miroslava Derenko, et al., "Whole mitochondrial genome diversity in two Hungarian populations," *Molecular Genetics and Genomics* 293, no. 5 (October 2018): supplemental data. This sample is coded as MG952865 in GenBank.

75 Agnieszka Piotrowska-Nowak, Ewa Kosior-Jarecka, et al., "Investigation of whole mitochondrial genome variation in normal tension glaucoma," *Experimental Eye Research* 178 (January 2019); Agnieszka Piotrowska-Nowak, Joanna L. Elson, et al., "New mtDNA Association Model, MutPred Variant Load, Suggests Individuals With Multiple Mildly Deleterious mtDNA Variants Are More Likely to Suffer From Atherosclerosis," *Frontiers in Genetics* 9 (2018); Boris A. Malyarchuk, Andrey Litvinov, et al., "Mitogenomic diversity in Russians and Poles," *Forensic Science International: Genetics* 30 (September 2017): supplemental data. These Polish samples are coded as MG646149, KY782163, KY782209, MH120449, and MN176243 in GenBank.

76 Miroslava Derenko, Boris A. Malyarchuk, et al., "Western Eurasian ancestry in modern Siberians based on mitogenomic data," *BMC Evolutionary Biology* 14 (October 10, 2014); Boris A. Malyarchuk, Miroslava Derenko, et al., "Mitogenomic diversity in Tatars from the Volga-Ural region of Russia," *Molecular Biology and Evolution* 27, no. 10 (2010); Malyarchuk, Litvinov, et al., "Mitogenomic diversity in Russians and Poles," supplemental data. These Russian samples are coded as KY670840, KJ856773, and GU122983 in GenBank.

77 Alissa Mittnik, Chuan-Chao Wang, et al., "The genetic prehistory of the Baltic Sea region," *Nature Communications* 9 (January 30, 2018): Table 1.

78 Översti, Majander, et al., "Human mitochondrial DNA lineages in Iron-Age Fennoscandia suggest incipient admixture and eastern introduction of farming-related maternal ancestry," supplementary data. The relevant sample is coded as MN540508 in GenBank.

79 This sample is coded as MK059589 in GenBank and has the alternative sample name LD24.

80 Mark Lipson, Anna Szécsényi-Nagy, et al., "Parallel palaeogenomic transects reveal complex genetic history of early European farmers," *Nature* 551, no. 7680 (November 16, 2017): Extended Data Table 1. This sample has the code names HAL39b and I2037.

81 Lipson, Szécsényi-Nagy, et al., "Parallel palaeogenomic transects...," Extended Data Table 1. This sample has the code names PULE1.23a and I2357.

82 Lipson, Szécsényi-Nagy, et al., "Parallel palaeogenomic transects...," Extended Data Table 1. This sample has the code names ESP30 and I0807.

83 "Jewish Ukraine West - mtDNA Test Results for Members."

84 Personal communication with Leo Cooper, May 6, 2021.

85 Helena Malmström, Anna Linderholm, et al., "Ancient mitochondrial DNA from the northern fringe of the Neolithic farming expansion in Europe sheds light on the dispersion process," *Philosophical Transactions of the Royal Society B: Biological Sciences* 370, no. 1660 (January 19, 2015): Table 1 where the sample is identified as "GE44 (KOP32) grave A5".

86 Översti, Majander, et al., "Human mitochondrial DNA lineages in Iron-Age Fennoscandia suggest incipient admixture and eastern introduction of farming-related maternal ancestry," supplementary data. The sample from Hollola, "Hollola12," is assigned the code MN540496 by GenBank and the one from Tuukkala, "Tuukkala8," is MN540518.

87 Ireneusz Stolarek, Anna Juras, et al., "A mosaic genetic structure of the human population living in the South Baltic region during the Iron Age," *Scientific Reports* 8 (February 6, 2018). They assigned this sample the identifier PCA0015.

88 Helena Malmström, Torsten Günther, et al., "The genomic ancestry of the Scandinavian Battle Axe Culture people and their relation to the broader Corded Ware horizon," *Proceedings of the Royal Society B: Biological Sciences* 286, no. 1912 (October 9, 2019): Table 1 on p. 3 where she is identified as sample "kar1".

89 Jeffrey D. Wexler, "mtDNA Ancestral Lines of Ashkenazi Jews," December 2018, accessed January 12, 2022, https://sites.google.com/view/ashkenazi-y-dna-and-mtdna/mtdna-ancestral-lines-of-ashkenazi-jews

90 "Greek DNA results," April 7, 2020 and September 13, 2020, accessed January 12, 2022, https://forums.gedmatch.com/BB/viewtopic.php?t=345&start=20

91   "The Haplogroup H&HV mtGenome Project: H1 - mtDNA Test Results for Members," Family Tree DNA, accessed January 13, 2022, https://www.familytreedna.com/public/mtDNA_H1?iframe=mtresults

92   This sample is coded as DQ523670 in GenBank.

93   Marina Silva, Gonzalo Oteo-García, et al., "Biomolecular insights into North African-related ancestry, mobility and diet in eleventh-century Al-Andalus," *Scientific Reports* 11 (September 13, 2021): supplemental data. They are coded as MZ920929 and MZ920286 in GenBank.

94   Kitti Maár, Gergely I. B. Varga, et al., "Maternal Lineages from 10–11th Century Commoner Cemeteries of the Carpathian Basin," *Genes (Basel)* 12, no. 3 (March 23, 2021): Table S1 where they are identified as samples "Ibrány-Esbóhalom/139" and "Homokmégy-Székes/50".

95   The relevant sample is coded as KY671070 in GenBank.

96   Kevin Alan Brook, "The Genetics of Crimean Karaites," *Karadeniz Araştırmaları* no. 42 (Summer 2014): 78 for sample codes K02 and K05.

97   "Karaites of E.Europe - mtDNA Test Results for Members," Family Tree DNA, accessed January 12, 2022, https://www.familytreedna.com/public/Karaite?iframe=mtresults

98   Costa, Pereira, et al., "A Substantial Prehistoric European Ancestry...," Figure 7, which only includes non-Ashkenazic carriers of this haplogroup.

99   Relevant samples are coded as KY409274, KY409362, and KY410129 in GenBank.

100 Óscar García, Santos Alonso, et al., "Forensically relevant phylogeographic evaluation of mitogenome variation in the Basque Country," *Forensic Science International: Genetics* 46 (May 2020). The relevant sample is coded as MN046430 in GenBank.

101 This sample is coded as MK059499 in GenBank and LP150 in Jennifer Klunk's research.

102 Personal communication with Leo Cooper, May 6, 2021.

103 A Jewish H1bo carrier, listing a matriline distantly from Spain and more recent ancestors from the Turkish city of Gallipoli and Greek city of Thessaloniki, participates in "Romaniote DNA Project - mtDNA Test Results for Members," Family Tree DNA, accessed January 13, 2022, https://www.familytreedna.com/public/romaniote?iframe=mtresults

104 "The Haplogroup H&HV mtGenome Project: H1 - mtDNA Test Results for Members." Several Ashkenazic H1bo carriers also belong to this project.

105 Personal communication with Ian Logan, June 7, 2021.

106 Ariel Toaff and Nadia Zeldes, "Taranto," in *Encyclopaedia Judaica*, 2nd ed., ed. Fred Skolnik (Detroit: Macmillan Reference USA, 2008), vol. 19.

107 Carla García-Fernández, Neus Font-Porterias et al., "Sex-biased patterns shaped the genetic history of Roma," *Scientific Reports* 10 (September 2, 2020): Table S3 where the Romanian H1bo is listed as RU038.

108 Costa, Pereira, et al., "A Substantial Prehistoric European Ancestry...," Figure 7, which only includes non-Ashkenazic carriers of this haplogroup.

109 Personal communication with Leo Cooper, May 6, 2021.

110 These samples are coded as KY408186 and KY409467 in GenBank.

111 The Upper Franconia sample is coded as MH000027 in GenBank.

112 "Alpine Y/mt-DNA - mtDNA Test Results for Members," Family Tree DNA, accessed January 13, 2022, https://www.familytreedna.com/public/

Alpine_DNA_Project_AlpGen_Genealogy?iframe=mtresults includes a non-Jewish H2a1e1a matriline descending from Catherine Blank, born about 1745 in Linden, Switzerland.

113 Doron M. Behar, Mannis van Oven, et al., "A 'Copernican' reassessment of the human mitochondrial DNA tree from its root," *American Journal of Human Genetics* 90, no. 4 (April 6, 2012): supplementary data. This sample is coded as JQ703366 in GenBank.

114 Shengting Li, Soren Besenbacher, et al., "Variation and association to diabetes in 2000 full mtDNA sequences mined from an exome study in a Danish population," *European Journal of Human Genetics* 22, no. 8 (August 2014); Nicola Raule, Federica Sevini, et al., "The co-occurrence of mtDNA mutations on different oxidative phosphorylation subunits, not detected by haplogroup analysis, affects human longevity and is population specific," *Aging Cell* 13, no. 3 (June 2014): supplementary data. These samples are coded as KF161175, KF162460, KF162877, JX152900, and JX152816 in GenBank.

115 Bernard Sécher, Rosa Fregel, et al., "The history of the North African mitochondrial DNA haplogroup U6 gene flow into the African, Eurasian and American continents," *BMC Evolutionary Biology* 14 (May 19, 2014): supplementary data where this sample has the identifier "TafVI-10".

116 The relevant sample has the code MT588230 in GenBank.

117 Personal communication with Leo Cooper, May 29, 2021.

118 Margaryan, Lawson, et al., "Population genomics of the Viking world." He is identified as sample VK512 in this study and also has the alternate identifier Estonia_Salme_II-Ü.

119 A relevant sample is coded as MK059526 in GenBank and LP208 in Jennifer Klunk's research.

120 A relevant sample is coded as MK059739 in GenBank and LM92 in Jennifer Klunk's research.

121 Pierre Zalloua, Catherine J. Collins, et al., "Ancient DNA of Phoenician remains indicates discontinuity in the settlement history of Ibiza," *Scientific Reports* 8 (December 4, 2018): Table 1. This sample's GenBank code is MH043583.

122 Gloria González-Fores, F. Tassi, et al., "A western route of prehistoric human migration from Africa into the Iberian Peninsula," *Proceedings of the Royal Society B: Biological Sciences* 286, no. 1895 (January 23, 2019). This sample is coded as LU339 in the article and MK321331 in GenBank.

123 Anja Furtwängler, Adam B. Rohrlach, et al., "Ancient genomes reveal social and genetic structure of Late Neolithic Switzerland," *Nature Communications* 11 (April 20, 2020): supplementary data. The relevant samples are coded as MT079020, MT079033, and MT079100 in GenBank.

124 Personal communication with Leo Cooper, June 8, 2021.

125 Gabriele Scorrano, Andrea Finocchio, et al., "The Genetic Landscape of Serbian Populations through Mitochondrial DNA Sequencing and Non-Recombining Region of the Y Chromosome Microsatellites," *Collegium Antropologicum* 41, no. 3 (2017): Tables 1 and 4.

126 Magdalena M. Buś, Maria Lembring, et al., "Mitochondrial DNA analysis of a Viking age mass grave in Sweden," *Forensic Science International: Genetics* 42 (September 2019): Table 1 on p. 270 and Table 3 on p. 273. This sample is coded as A27 in the article.

127 "Jewish Ukraine West - mtDNA Test Results for Members."

128 Behar, van Oven, et al., "A 'Copernican' reassessment of the human mitochondrial DNA tree from its root," supplementary data. The relevant sample is coded as JQ703544 in GenBank.

129 Personal communication with Leo Cooper, May 29, 2021.

130 These samples are coded as FJ460550, FJ460553, and FJ460556 in GenBank.

131 Personal communication with Stuart Drucker, March 29, 2021.

132 Furtwängler, Rohrlach, et al., "Ancient genomes reveal social and genetic structure of Late Neolithic Switzerland," supplementary data. The relevant sample is coded as MT079053 in GenBank.

133 Patterson, Isakov, et al., "Large-scale migration into Britain...," Supplementary Table 1 listing her sample code I12608.

134 This sample is coded as MK059504 in GenBank and has the alternative identifier LP161.

135 Lara M. Cassidy, Ros Ó Maoldúin, et al., "A dynastic elite in monumental Neolithic society," *Nature* 582, no. 7812 (June 17, 2020): supplementary data where he has the identifiers PB768 and Parknabinnia768.

136 "Jewish Ukraine West - mtDNA Test Results for Members."

137 Personal communication with Leo Cooper, May 29, 2021.

138 This sample is coded as MK059672 in GenBank and LG30 in Jennifer Klunk's research.

139 This sample is coded as MK059491 in GenBank and LP136 in Jennifer Klunk's research.

140 Margaryan, Lawson, et al., "Population genomics of the Viking world." He is identified as sample VK399 in this study.

141 Maár, Varga, et al., "Maternal Lineages from 10–11th Century Commoner Cemeteries of the Carpathian Basin," Table S1 where they are identified as sample 17 from Sárrétudvari-Hízóföld, sample 327 from Püspökladány-Eperjesvölgy, and sample 15 from Magyarhomorog-Kónyadomb.

142 Corina Knipper, Alissa Mittnik, et al., "Female exogamy and gene pool diversification at the transition from the Final Neolithic to the Early Bronze Age in central Europe," *Proceedings of the National Academy of Sciences of the United States of America* 114, no. 38 (September 19, 2017): supplementary data. This sample has the code MF498663 in GenBank and also has the alternative identifier WEHR_1193.

143 Iñigo Olalde, Selina Brace, et al., "The Beaker phenomenon and the genomic transformation of northwest Europe," *Nature* 555, no. 7695 (March 8, 2018): Supplementary Table 1. This sample was given the identifier I4884.

144 Wolfgang Haak, Iosif Lazaridis, et al., "Massive migration from the steppe was a source for Indo-European languages in Europe," *Nature* 522, no. 7555 (June 11, 2015): Supplementary Data 1 where she is assigned the identifiers I0803 and EUL41.

145 Cassidy, Ó Maoldúin, et al., "A dynastic elite in monumental Neolithic society," supplementary data where she has the identifiers PN113 and Poulnabrone10_113.

146 Personal communication with Leo Cooper, May 29, 2021.

147 "The Haplogroup H&HV mtGenome Project: H4 - mtDNA Test Results for Members," Family Tree DNA, accessed January 13, 2022, https://www.familytreedna.com/public/mtDNA_H4?iframe=mtresults

148 "The Cape Dutch DNA Stamouer Project - mtDNA Test Results for Members," Family Tree DNA, accessed January 13, 2022, https://www.familytreedna.com/public/CapeDutch?iframe=mtresults

149 Maja Krzewińska, Anna Kjellström, et al., "Genomic and Strontium Isotope Variation Reveal Immigration Patterns in a Viking Age Town," *Current Biology* 28 (September 10, 2018): Table 1 on p. 2732 where she has the identifier "bns023".

150 Behar, van Oven, et al., "A 'Copernican' reassessment of the human mitochondrial DNA tree from its root," supplementary data. This sample is coded in GenBank as JQ705703.

151 Personal communication with Martin Davis, May 16, 2016.

152 "Jewish Ukraine West - mtDNA Test Results for Members."

153 Personal communication with Jeffrey Wexler, April 7, 2021.

154 Dick Wåhlin, "Eventuellt så kan jag vara släkt med några av de flyktingar som just nu flyr från Syrien," April 9, 2016, accessed September 9, 2021, https://dickwahlin. wordpress.com/2016/04/09/eventuellt-sa-kanske-jag-ar-slakt-med-nagra-av-de-flyktingar-som-nu-flyr-fran-syrien/

155 Aurelia Santoro, Valentina Balbi, et al., "Evidence for Sub-Haplogroup H5 of Mitochondrial DNA as a Risk Factor for Late Onset Alzheimer's Disease," *PLoS ONE* 5, no. 8 (August 6, 2010). Inside GenBank's database, the H5r* Italians from this study have the codes GQ983083, GQ983085, GQ983086, GQ983094, and GQ983107, the H5r1 Italian is GQ983103, the H5r2 Italians are GQ983061, GQ983062, and GQ983097, the H5s* Italian is GQ983108, and the H5t Italian is GQ983098.

156 Doron M. Behar, Christine Harmant, et al., "The Basque Paradigm: Genetic Evidence of a Maternal Continuity in the Franco-Cantabrian Region since Pre-Neolithic Times," *American Journal of Human Genetics* 90, no. 3 (March 9, 2012). The relevant sample is coded as JQ324596 in GenBank.

157 Maár, Varga, et al., "Maternal Lineages from 10–11th Century Commoner Cemeteries of the Carpathian Basin," Table S1 where he is identified as sample "Homokmégy-Székes/49".

158 Anna Juras, Miroslawa Dabert, et al., "Ancient DNA Reveals Matrilineal Continuity in Present-Day Poland over the Last Two Millennia," *PLoS ONE* 9, no. 10 (October 22, 2014): "Discussion" section.

159 The relevant samples are coded as MT588271 and MT588278 in GenBank.

160 Lehti Saag, Liivi Varul, et al., "Extensive Farming in Estonia Started through a Sex-Biased Migration from the Steppe," *Current Biology* 27, no. 14 (July 24, 2017): Table 1 on p. 2187. This sample has the codes RISE00 in this study and MG429029 in GenBank.

161 Olalde, Brace, et al., "The Beaker phenomenon...," Supplementary Table 1. The Bavarian samples were assigned the identifiers I4249 (RISE917) and I5655 (RISE919) and the North Holland sample is I4075.

162 Luísa Pereira, Martin Richards, et al., "High-resolution mtDNA evidence for the late-glacial resettlement of Europe from an Iberian refugium," *Genome Research* 15, no. 1 (January 2005): especially "Results" section. They tallied their European H5a samples in Table 1.

163 Christine Keyser, Caroline Bouakaze, et al., "Ancient DNA provides new insights into the history of south Siberian Kurgan people," *Human Genetics* 126 (2009): Table 1 where he is identified as sample S32.

164 Keyser, Bouakaze, et al., "Ancient DNA provides new insights into the history of south Siberian Kurgan people," 404 and Table 5.

165 "Jewish Ukraine West - mtDNA Test Results for Members."

166   Behar, van Oven, et al., "A 'Copernican' reassessment of the human mitochondrial DNA tree from its root," supplementary data. Its code is JQ704150 in GenBank.

167   This sample's code is KF161664 in GenBank.

168   This sample's code is HQ384189 in GenBank.

169   Personal communication with Leo Cooper, May 29, 2021.

170   The relevant sample is coded as MK059502 in GenBank and LP157 in Jennifer Klunk's research.

171   "German Jewish Gersig DNA Project - mtDNA Test Results for Members," Family Tree DNA, accessed March 26, 2021, https://www.familytreedna.com/public/GermanJewishGersig?iframe=mtresults

172   Personal communication with Leo Cooper, May 29, 2021.

173   "Telsiai Uyezd/District, Lithuania Jewish Geography Project - mtDNA Test Results for Members," Family Tree DNA, accessed January 15, 2022, https://www.familytreedna.com/public/telsiaiuyezd/default.aspx?section=mtresults

174   Personal communication with Leo Cooper, March 27 and May 31, 2021.

175   This person, Felipe González de Otoya, wrote this to "mtDNA Haplogroup H - Subgroup H6," Facebook, accessed January 15, 2022, https://www.facebook.com/groups/48909723061/permalink/10155166598918062/ and another participant in that conversation is Yeho Erz whose H6a1a1a matriline is "Sefaradic Jewish from Smyrna."

176   Costa, Pereira, et al., "A Substantial Prehistoric European Ancestry...," Figure 8 on p. 7.

177   Ibid., Supplementary Table S7.

178   Olalde, Brace, et al., "The Beaker phenomenon...," Supplementary Table 1. She was assigned the identifier I4071.

179   Peter de Barros Damgaard, Nina Marchi, et al., "137 ancient human genomes from across the Eurasian steppes," *Nature* 557, no. 7705 (May 9, 2018): supplementary data. She was assigned the identifier DA121.

180   This sample's code is HQ730608 in GenBank.

181   Finland samples in GenBank include KY606236 (an ethnic Finn) and JQ703028.

182   "The Haplogroup H&HV mtGenome Project: H6 - mtDNA Test Results for Members," Family Tree DNA, accessed January 16, 2022, https://www.familytreedna.com/public/mtDNA_H6?iframe=mtresults

183   This sample is coded as MK059651 in GenBank and LE16 in Jennifer Klunk's research.

184   "Jewish Ukraine West - mtDNA Test Results for Members."

185   Margaryan, Lawson, et al., "Population genomics of the Viking world." He is identified as sample VK327 in this study.

186   The relevant sample is coded as MK059627 in GenBank and LE114 in Jennifer Klunk's research.

187   Rui Martiniano, Anwen Caffell, et al., "Genomic signals of migration and continuity in Britain before the Anglo-Saxons," *Nature Communications* 7 (January 19, 2016): Table 1 where he is identified as sample 6DRIF-23.

188   Patterson, Isakov, et al., "Large-scale migration into Britain...," Supplementary Table 1 listing his sample code I6769.

189   Olalde, Brace, et al., "The Beaker phenomenon...," Supplementary Table 1. He was assigned the identifiers I2655 and GENSCOT24.

190   Haak, Lazaridis, et al., "Massive migration from the steppe...," Extended Data Table 2; Olalde, Brace, et al., "The Beaker phenomenon...," Supplementary Table

1; Morten E. Allentoft, Martin Sikora, et al., "Population genomics of Bronze Age Eurasia," *Nature* 522, no. 7555 (June 11, 2015): Supplementary Table 14.

191 Marta Mielnik-Sikorska, Patrycja Daca, et al., "The History of Slavs Inferred from Complete Mitochondrial Genome Sequences," *PLoS ONE* 8, no. 1 (January 14, 2013).

192 Malmström, Günther, et al., "The genomic ancestry of the Scandinavian...," Table 1 where he is identified as sample oll009.

193 Corina Knipper, Matthias Fragata, et al., "A distinct section of the Early Bronze Age society? Stable isotope investigations of burials in settlement pits and multiple inhumations of the Únětice culture in central Germany," *American Journal of Physical Anthropology* 159, no. 3 (March 2016) where they are identified as samples LEA 1 and LEA 4.

194 Personal communication with Leo Cooper, March 25, 2021.

195 Raule, Sevini, et al., "The co-occurrence of mtDNA mutations...," supplementary data. This sample is coded as JX152999 in GenBank.

196 Anna Olivieri, Carlo Sidore, et al., "Mitogenome Diversity in Sardinians: A Genetic Window onto an Island's Past," *Molecular Biology and Evolution* 34, no. 5 (May 2017). This sample is coded as KY409915 in GenBank.

197 Iain Mathieson, Songül Alpaslan-Roodenberg, et al., "The genomic history of southeastern Europe," *Nature* 555, no. 7695 (March 8, 2018): Supplementary Information where he is identified as sample I5072.

198 Lipson, Szécsényi-Nagy, et al., "Parallel palaeogenomic transects...," Extended Data Table 1 where he is identified as sample TISO1b.

199 Mathieson, Alpaslan-Roodenberg, et al., "The genomic history of southeastern Europe," Supplementary Information where she is identified as sample I0785.

200 Mario Novak, Iñigo Olalde, et al., "Genome-wide analysis of nearly all the victims of a 6200 year old massacre," *PLoS ONE* 16, no. 3 (March 10, 2021). She has been assigned the identifier I10057.

201 Doron Yacobi and Felice L. Bedford, "Evidence of Early Gene Flow between Ashkenazi Jews and Non-Jewish Europeans in Mitochondrial DNA Haplogroup H7," *Journal of Genetic Genealogy* 8, no. 1 (Fall 2016): 26. Most of the people in this sentence were discussed by Yacobi and Bedford but at the time of their study they found only two of the four current Germans. The person from Albania is only in YFull's MTree.

202 Personal communication with Leo Cooper, June 8, 2021.

203 Yacobi and Bedford, "Evidence of Early Gene Flow...," 26, 28.

204 Ibid., 28.

205 Olivieri, Sidore, et al., "Mitogenome Diversity in Sardinians." This sample is coded as KY409861 in GenBank.

206 Guido A. Gnecchi-Ruscone, Elmira Khussainova, et al., "Ancient genomic time transect from the Central Asian Steppe unravels the history of the Scythians," *Science Advances* 7, no. 13 (March 26, 2021): "Supplementary Materials" and supplementary data where this sample has the identifier "BIY001.A".

207 GenBank includes the relevant samples KX788167 and KX789084.

208 Yacobi and Bedford, "Evidence of Early Gene Flow...," 21, 23, 29.

209 GenBank includes the relevant samples JQ703313, KX679399, KX690095, KX702230 (mine), KX784496, and KY077676.

210 Personal communication with Felice Bedford, May 31, 2021.

211 Li, Besenbacher, et al., "Variation and association to diabetes in 2000 full mtDNA sequences mined from an exome study in a Danish population." This sample is coded as KF162452 in GenBank and confirmed to carry the key mutations T1700C and T11137C, among other mutations that the definite Ashkenazic H7j1 carriers share. The study's 2,000 samples, including this one, derive from the medically oriented study Anders Albrechtsen, N. Grarup, et al., "Exome sequencing-driven discovery of coding polymorphisms associated with common metabolic phenotypes," *Diabetologia* 56, no. 2 (2013), "Introduction" section, which called those participants "Danish individuals."

212 "Jewish Ukraine West - mtDNA Test Results for Members."

213 Reyhan Yaka, Ayşegül Birand, et al., "Archaeogenetics of Late Iron Age Çemialo Sırtı, Batman: Investigating maternal genetic continuity in north Mesopotamia since the Neolithic," *American Journal of Physical Anthropology* 166, no. 1 (May 2018). This sample has been given the code name SK23.

214 Personal communication with Leo Cooper, May 6, 2021.

215 Ioana Rusu, Alessandra Modi, et al., "Mitochondrial ancestry of medieval individuals carelessly interred in a multiple burial from southeastern Romania," *Scientific Reports* 9 (January 30, 2019).

216 Översti, Majander, et al., "Human mitochondrial DNA lineages in Iron-Age Fennoscandia suggest incipient admixture and eastern introduction of farming-related maternal ancestry," supplementary data. The relevant sample is coded as MN540516 in GenBank.

217 "Jewish Ukraine West - mtDNA Test Results for Members."

218 GenBank includes Finnish H11a2a2 carriers (MN516558 and JX153593), Polish carriers (KY782226 and MH120705), and carriers from Russia's Tula region (KY670995, KY671049, and KY671059).

219 Personal communication with Leo Cooper, May 6, 2021.

220 This sample is coded as MK059657 in GenBank and LG06 in Jennifer Klunk's research.

221 Mittnik, Wang, et al., "The genetic prehistory of the Baltic Sea region," Table 1.

222 "Jewish Ukraine West - mtDNA Test Results for Members."

223 Personal communication with Leo Cooper, May 6, 2021.

224 A relevant sample is coded as KF161213 in GenBank.

225 Relevant samples are coded as MG646200, MH120498, MH120654, and MH120816 in GenBank.

226 A relevant sample is coded as MN610428 in GenBank.

227 Mari Järve, Lehti Saag, et al., "Shifts in the Genetic Landscape of the Western Eurasian Steppe Associated with the Beginning and End of the Scythian Dominance," *Current Biology* 29, no. 14 (July 22, 2019): Table S2.

228 Personal communication with Jeffrey Wexler, May 15, 2021.

229 Personal communication with Leo Cooper, May 6, 2021.

230 Personal communication with Jeffrey Wexler, June 4, 2021.

231 Patterson, Isakov, et al., "Large-scale migration into Britain...," Supplementary Table 1 listing the relevant samples I13726, I16591, I16616, and I19872.

232 Iñigo Olalde, Swapan Mallick, et al., "The genomic history of the Iberian Peninsula over the past 8000 years," *Science* 363, no. 6432 (2019): supplementary data. She was assigned the identifiers I8203 and "02-SU-33-A4-T1058".

233 Haak, Lazaridis, et al., "Massive migration from the steppe...," Supplementary Data 1 where he is assigned the identifiers I0370 and SVP10.

234 A relevant sample is coded as MH120572 in GenBank.

235 A relevant sample is coded as MK134336 in GenBank.

236 Personal communication with Leo Cooper, May 6, 2021.

237 This sample is coded as MK059703 in GenBank and LM40 in Jennifer Klunk's research.

238 Lehti Saag, Margot Laneman, et al., "The Arrival of Siberian Ancestry Connecting the Eastern Baltic to Uralic Speakers further East," *Current Biology* 29, no. 10 (May 20, 2019): Table 1 where he has the sample code 0LS10.

239 Modern Danish examples of H15b include GenBank samples JX153635, KF161667, KF162091, KF162847, and KF162889.

240 Examples of H15b GenBank samples are AY738960, EU600353, MF362827, and MK134270. Other H15b carriers are listed in "The Haplogroup H&HV mtGenome Project: H15 - mtDNA Test Results for Members," Family Tree DNA, accessed January 15, 2022, https://www.familytreedna.com/public/mtDNA_H15?iframe=mtresults

241 The relevant sample is coded as MK059529 in GenBank and LP212 in Jennifer Klunk's research.

242 Costa, Pereira, et al., "A Substantial Prehistoric European Ancestry...," Supplementary Table S7. This Bulgarian Jew has the sample number EF556189 in GenBank.

243 Personal communication with Leo Cooper, May 29, 2021.

244 Costa, Pereira, et al., "A Substantial Prehistoric European Ancestry...," Supplementary Note 3, p. 38.

245 Personal communication with Leo Cooper, May 31, 2021.

246 Lipson, Szécsényi-Nagy, et al., "Parallel palaeogenomic transects...," Extended Data Table 1.

247 Ibid. She is identified as sample HAL32b.

248 Novak, Olalde, et al., "Genome-wide analysis of nearly all the victims of a 6200 year old massacre." She has been assigned the identifier I10071.

249 Lipson, Szécsényi-Nagy, et al., "Parallel palaeogenomic transects...," Extended Data Table 1.

250 "Jewish Ukraine West - mtDNA Test Results for Members."

251 Behar, van Oven, et al., "A 'Copernican' reassessment of the human mitochondrial DNA tree from its root," supplementary data. This sample is coded as JQ702924 in GenBank.

252 Veronika Csáky, Dániel Gerber, et al., "Early medieval genetic data from Ural region evaluated in the light of archaeological evidence of ancient Hungarians," *Scientific Reports* 10 (November 5, 2020): especially Supplementary Information: Table S16 and Figure S4g.

253 Lily Agranat-Tamir, Shamam Waldman, et al., "The Genomic History of the Bronze Age Southern Levant," *Cell* 181, no. 5 (May 28, 2020): supplemental data. It has the sample IDs I6928 and Levant47.

254 Vagheesh M. Narasimhan, Nick Patterson, et al., "The formation of human populations in South and Central Asia," *Science* 365, no. 6457 (September 6, 2019): supplemental data. This sample has been assigned the identifier I11469.

255 This sample is coded as MK059670 in GenBank and has the alternative code LG28.

256 Haak, Lazaridis, et al., "Massive migration from the steppe...," Supplementary Data 1 where it is assigned the identifiers I0027 and LBK2172.

257 Lipson, Szécsényi-Nagy, et al., "Parallel palaeogenomic transects...," Extended Data Table 1. This sample has the code names VEJ9a and I2394.

258 Olalde, Brace, et al., "The Beaker phenomenon...," Supplementary Table 1. She was assigned the identifiers I6538 and HB0057.
259 Mathieson, Alpaslan-Roodenberg, et al., "The genomic history of southeastern Europe," Supplementary Information where he is identified as sample I4666.
260 "Jewish Ukraine West - mtDNA Test Results for Members."
261 Personal communication with Leo Cooper, May 31, 2021.
262 Buś, Lembring, et al., "Mitochondrial DNA analysis of a Viking age mass grave in Sweden," Table 1 on p. 270 and Table 3 on p. 273. These samples are coded as A4 and A9 in the article.
263 Knipper, Mittnik, et al., "Female exogamy and gene pool diversification at the transition from the Final Neolithic to the Early Bronze Age in central Europe," supplementary data. This sample has the code MF498697 in GenBank and also has the alternative identifier POST 35.
264 "Jewish Ukraine West - mtDNA Test Results for Members."
265 Personal communication with Jonah Stern, April 14, 2019.
266 "The Mitochondrial DNA (mtDNA) Haplogroup H & HV Project - mtDNA Test Results for Members," Family Tree DNA, accessed October 4, 2018, https://www.familytreedna.com/public/H%20mtDNA%20Haplogroup?iframe=mtresults
267 A relevant sample is coded as MN977122 in GenBank.
268 A relevant sample is coded as MK617262 in GenBank.
269 Maár, Varga, et al., "Maternal Lineages from 10–11th Century Commoner Cemeteries of the Carpathian Basin," Table S1 where they are identified as samples 118 and 122 from Vörs-Papkert-B.
270 Eirini Skourtanioti, Yilmaz S. Erdal, et al., "Genomic History of Neolithic to Bronze Age Anatolia, Northern Levant, and Southern Caucasus," *Cell* 181, no. 5 (May 28, 2020). This sample has the identifier ALA013 in Table S9.
271 Matthew V. Emery, Ana T. Duggan, et al., "Ancient Roman mitochondrial genomes and isotopes reveal relationships and geographic origins at the local and pan-Mediterranean scales," *Journal of Archaeological Science: Reports* 20 (August 2018): supplementary data. The relevant sample is coded as MG773654 in GenBank and also carries the identifiers LRV 135 and F287.
272 Personal communication with Leo Cooper, May 29, 2021.
273 "Jewish Ukraine West - mtDNA Test Results for Members."
274 Behar, van Oven, et al., "A 'Copernican' reassessment of the human mitochondrial DNA tree from its root," supplementary data. This sample is coded as JQ703478 in GenBank.
275 This sample is coded as MH120478 in GenBank.
276 David W. Collins, Harini V. Gudiseva, et al., "Association of primary open-angle glaucoma with mitochondrial variants and haplogroups common in African Americans," *Molecular Vision* 22 (May 16, 2016): Figure 4, accessed October 4, 2021, https://www.ncbi.nlm.nih.gov/pmc/articles/PMC4872278/figure/f4/
277 Justyna Jarczak, Łukasz Grochowalski, et al., "Mitochondrial DNA variability of the Polish population," *European Journal of Human Genetics* 27 (March 21, 2019): Table S3.
278 The Sicilian sample is coded as KY399172 in GenBank.
279 Personal communication with Anthony Lizza, April 19, 2021.
280 Joana F. Ferragut, Cristian Ramon, et al., "Middle Eastern genetic legacy in the paternal and maternal gene pools of Chuetas," *Scientific Reports* 10 (December 8, 2020): Table 3.

281 Personal communication with Leo Cooper, May 29, 2021.

282 Inês Nogueiro, João C. Teixeira, et al., "Portuguese crypto-Jews: the genetic heritage of a complex history," *Frontiers in Genetics* 6 (February 2, 2015): 8.

283 A relevant sample is coded as MN046464 in GenBank.

284 Costa, Pereira, et al., "A Substantial Prehistoric European Ancestry..."

285 Doron M. Behar, Ene Metspalu, et al., "Counting the Founders: The Matrilineal Genetic Ancestry of the Jewish Diaspora," *PLoS ONE* 3, no. 4 (April 30, 2008).

286 Lara M. Cassidy, Rui Martiniano, et al., "Neolithic and Bronze Age migration to Ireland and establishment of the insular Atlantic genome," *Proceedings of the National Academic of Sciences of the United States of America* 113, no. 2 (January 12, 2016): p. 25 of "Appendix S7: Mitochondrial Genome Analysis". Her sample carries the identifier BA64.

287 These samples are coded as KY797184 and MK059626 respectively in GenBank.

288 Personal communication with Leo Cooper, May 31, 2021.

289 Sara De Fanti, Chiara Barbieri, et al., "Fine Dissection of Human Mitochondrial DNA Haplogroup HV Lineages Reveals Paleolithic Signatures from European Glacial Refugia," *PLoS ONE* 10, no. 12 (December 7, 2015): S1 Table. Gasparre's Italian is coded as EF660936 in GenBank. The Italian from Aviano carries the code DB2192 in De Fanti's database and KP340155 in GenBank while the Italian from Vicenza carries the codes DB532 and KP340156.

290 The Palestinian samples are coded as HGDP00685 and HGDP00729 in the Human Genome Diversity Project.

291 Alessandra Modi, Desislava Nesheva, et al., "Ancient human mitochondrial genomes from Bronze Age Bulgaria: new insights into the genetic history of Thracians," *Scientific Reports* 9 (April 1, 2019): "Methods" section, Table 1, and Table 2 where this sample is identified as "SM 24.2".

292 Verena J. Schuenemann, Alexander Peltzer, et al., "Ancient Egyptian mummy genomes suggest an increase of Sub-Saharan African ancestry in post-Roman periods," *Nature Communications* 8 (May 30, 2017): supplemental data where the Roman-era male carries the identifier JK2907 and the Ptolemaic samples carry the identifiers JK2985 and JK2987.

293 Maria Angela Diroma, Alessandra Modi, et al., "New Insights Into Mitochondrial DNA Reconstruction and Variant Detection in Ancient Samples," *Frontiers in Genetics* 12 (February 18, 2021): Figure 2. The relevant sample is coded as MW389248 in GenBank.

294 Éadaoin Harney, Hila May, et al., "Ancient DNA from Chalcolithic Israel reveals the role of population mixture in cultural transformation," *Nature Communications* 9 (August 20, 2018): Supplementary Data 1 where this sample is given the identifier I1165.

295 "HV1b MtDNA Match Group - mtDNA Test Results for Members," Family Tree DNA, accessed January 13, 2022, https://www.familytreedna.com/public/HV1dkfgsgang?iframe=mtresults

296 "Romaniote DNA Project - mtDNA Test Results for Members" lists this person's matrilineal ancestor as Behora Cohen from the nineteenth century.

297 The Armenian sample is coded as MK491372 in GenBank.

298 "Today my first results from 23andMe," October 12, 2013, accessed February 27, 2022, https://www.eupedia.com/forum/archive/index.php/t-29151.html

299 Katherine S. Elliott, Marc Haber, et al., "Fine-scale genetic structure in the United Arab Emirates reflects endogamous and consanguineous culture, population history and geography," *Molecular Biology and Evolution* 39, no. 3 (March 2022): Supplementary Data where the sample number is 2211.

300 Personal communication with Anthony Lizza, April 19, 2021.

301 Schuenemann, Peltzer, et al., "Ancient Egyptian mummy genomes suggest an increase of Sub-Saharan African ancestry in post-Roman periods," supplemental data where this sample carries the identifier JK2896.

302 Costa, Pereira, et al., "A Substantial Prehistoric European Ancestry...," Supplementary Note 3, p. 42.

303 Michel Shamoon-Pour, Mian Li, et al., "Rare human mitochondrial HV lineages spread from the Near East and Caucasus during post-LGM and Neolithic expansions," *Scientific Reports* 9 (October 14, 2019): "Results and Discussion" section.

304 Skourtanioti, Erdal, et al., "Genomic History of Neolithic to Bronze Age Anatolia, Northern Levant, and Southern Caucasus." This sample has the identifier ALA018 in Table S9.

305 Agranat-Tamir, Waldman, et al., "The Genomic History of the Bronze Age Southern Levant," supplemental data. It has the sample IDs I3965 and Levant37.

306 Eliška Musilová, Verónica Fernandes, et al., "Population History of the Red Sea— Genetic Exchanges Between the Arabian Peninsula and East Africa Signaled in the Mitochondrial DNA HV1 Haplogroup," *American Journal of Physical Anthropology* 145 (2011): 593.

307 "HV1b MtDNA Match Group - mtDNA Test Results for Members" includes an Ashkenazic HV5a matriline from Latvia. Ashkenazic HV5a carriers in GenBank include samples EF419890 (matriline from Latvia), DQ377992 and JQ703615 (matrilines from Belarus), and JN034636 and EU558385 (matrilines from Poland).

308 Anna Schönberg, Christoph Theunert, et al., "High-throughput sequencing of complete human mtDNA genomes from the Caucasus and West Asia: high diversity and demographic inferences," *European Journal of Human Genetics* 19 (April 13, 2011): supplementary data. This study's Armenian sample is coded as HM852778 in GenBank.

309 An Indian HV5a carrier is coded as JQ446397 in GenBank.

310 "Jewish Ukraine West - mtDNA Test Results for Members."

311 Behar, van Oven, et al., "A 'Copernican' reassessment of the human mitochondrial DNA tree from its root," supplementary data. This sample is coded as JQ702655 in GenBank.

312 Anna Olivieri, Maria Pala, et al., "Mitogenomes from Two Uncommon Haplogroups Mark Late Glacial/Postglacial Expansions from the Near East and Neolithic Dispersals within Europe," *PLoS ONE* 8, no. 7 (July 31, 2013): "Discussion" section.

313 Costa, Pereira, et al., "A Substantial Prehistoric European Ancestry...," Supplementary Figures.

314 Olivieri, Pala, et al., "Mitogenomes from Two Uncommon Haplogroups...," "Discussion" section.

315 Marc Haber, Joyce Nassar, et al., "A Genetic History of the Near East from an aDNA Time Course Sampling Eight Points in the Past 4,000 Years," *American Journal of Human Genetics* 107, no. 1 (July 2, 2020): Tables S1 and S4 where he is identified as sample SFI-15.

316 Gnecchi-Ruscone, Khussainova, et al., "Ancient genomic time transect from the Central Asian Steppe unravels the history of the Scythians," "Supplementary Materials" and supplementary data where this sample has the identifier "KNT001.A".

317 "mtDNA I5a1," Anthrogenica, post #9 dated September 4, 2018, accessed January 11, 2022, https://anthrogenica.com/showthread.php?3549-mtDNA-I5a1

318 Mittnik, Wang, et al., "The genetic prehistory of the Baltic Sea region," Table 1. This sample is coded as MG429043 in GenBank.

319 Allentoft, Sikora, et al., "Population genomics of Bronze Age Eurasia," supplemental data where this sample is coded as RISE435.

320 Olalde, Brace, et al., "The Beaker phenomenon...," Supplementary Table 1. They were assigned the identifiers I5521 and I5522.

321 Damgaard, Marchi, et al., "137 ancient human genomes from across the Eurasian steppes," supplementary data. He was assigned the identifier DA221.

322 Costa, Pereira, et al., "A Substantial Prehistoric European Ancestry...," "Results" section.

323 Maria Pala, Anna Olivieri, et al., "Mitochondrial DNA Signals of Late Glacial Recolonization of Europe from Near Eastern Refugia," *American Journal of Human Genetics* 90, no. 5 (May 4, 2012): supplementary data. The Greek samples are coded as JQ797847, JQ797848, JQ797851, and JQ797854 in GenBank.

324 Personal communication with Anthony Lizza, April 19, 2021.

325 This sample is coded as KY782162 in GenBank.

326 Florian Clemente, Martina Unterländer, et al., "The genomic history of the Aegean palatial civilizations," *Cell* 184, no. 10 (May 13, 2021): Table 2 where she is listed as sample Log04.

327 Elizabeth Matisoo-Smith, Anna L. Gosling, et al., "Ancient mitogenomes of Phoenicians from Sardinia and Lebanon: A story of settlement, integration, and female mobility," *PLoS ONE* 13, no. 1 (January 10, 2018): supplemental data. The relevant sample is coded as KY797213 in GenBank and has the alternative identifier LEB52_32AR21.

328 Personal communication with Anthony Lizza, April 19, 2021.

329 These samples are coded as MH542434 and MH120526 in GenBank.

330 This sample is coded as KX440214 in GenBank.

331 The relevant sample is coded as MK059545 in GenBank and LP240 in Jennifer Klunk's research.

332 Olalde, Brace, et al., "The Beaker phenomenon...," Supplementary Table 1. He was assigned the identifiers I2691 and GENSCOT30.

333 Lehti Saag, Sergey V. Vasilyev, et al., "Genetic ancestry changes in Stone to Bronze Age transition in the East European plain," *Science Advances* 7, no. 4 (January 20, 2021): Table 1 where they are assigned the identifiers BOL002, NIK005, and VOD001.

334 Olalde, Brace, et al., "The Beaker phenomenon...," Supplementary Table 1. He was assigned the identifier I0457.

335 Iain Mathieson, Iosif Lazaridis, et al., "Genome-wide patterns of selection in 230 ancient Eurasians," *Nature* 528, no. 7583 (November 23, 2015): supplemental data. She has the identifiers I1280 and MIR17.

336 Olalde, Mallick, et al., "The genomic history of the Iberian Peninsula over the past 8000 years," supplementary data. He was assigned the identifiers I10280 and "GN.89.E2.379a".

337 Olalde, Mallick, et al., "The genomic history of the Iberian Peninsula over the past 8000 years," supplementary data. He was assigned the identifier I7602.

338 Lipson, Szécsényi-Nagy, et al., "Parallel palaeogenomic transects...," Extended Data Table 1. They were assigned the identifiers I1975 and I1981 respectively.

339 Olalde, Mallick, et al., "The genomic history of the Iberian Peninsula over the past 8000 years," supplementary data. He was assigned the identifier I12877.

340 Zuzana Hofmanova, Susanne Kreutzer, et al., "Early farmers from across Europe directly descended from Neolithic Aegeans," *Proceedings of the National Academy of Sciences of the United States of America* 113, no. 25 (2016): supplementary data. She was assigned the identifier Pal7.

341 Mathieson, Alpaslan-Roodenberg, et al., "The genomic history of southeastern Europe," Supplementary Information. He was assigned the identifiers I0676, MACE7, "KV VIII", and "SE 513".

342 Cristina Gamba, Eppie R. Jones, et al., "Genome flux and stasis in a five millennium transect of European prehistory," *Nature Communications* 5 (October 21, 2014): supplementary data. He was assigned the identifiers I1500, HUNG372, and NE5.

343 Olivieri, Sidore, et al., "Mitogenome Diversity in Sardinians." This Sardinian sample is coded as KY409278 in GenBank.

344 Pala, Olivieri, et al., "Mitochondrial DNA Signals of Late Glacial Recolonization of Europe from Near Eastern Refugia," supplementary data. This Italian sample is coded as JQ797827 in GenBank.

345 Martiniano, Caffell, et al., "Genomic signals of migration and continuity in Britain before the Anglo-Saxons," Table 1 where he is identified as sample 6DRIF-21.

346 Gundula Müldner, Carolyn Chenery, and Hella Eckardt, "The 'Headless Romans': Multi-isotope investigations of an unusual burial ground from Roman Britain," *Journal of Archaeological Science* 38, no. 2 (February 2011): section 5.1: "Oxygen and Strontium Isotopes."

347 Alessandra Modi, Hovirag Lancioni, et al., "The mitogenome portrait of Umbria in Central Italy as depicted by contemporary inhabitants and pre-Roman remains," *Scientific Reports* 10 (July 1, 2020). This sample is coded as MN687306 in GenBank.

348 Costa, Pereira, et al., "A Substantial Prehistoric European Ancestry...," "Results" section.

349 Pala, Olivieri, et al., "Mitochondrial DNA Signals of Late Glacial Recolonization of Europe from Near Eastern Refugia," Table S1.

350 Haak, Lazaridis, et al., "Massive migration from the steppe...," Supplementary Data 1 where she is assigned the identifiers I0113 and QUEXII4.

351 Mathieson, Lazaridis, et al., "Genome-wide patterns of selection in 230 ancient Eurasians," supplemental data. One has the identifiers I1538 and ESP20 while the other has I1540 and ESP28.

352 Gamba, Jones, et al., "Genome flux and stasis in a five millennium transect of European prehistory," supplementary data. She was assigned the identifiers I1505, PF839/1198, and NE4.

353 Mathieson, Alpaslan-Roodenberg, et al., "The genomic history of southeastern Europe," Supplementary Information where she is identified as sample I2533.

354 Federico Sánchez-Quinto, Helena Malmström, et al., "Megalithic tombs in western and northern Neolithic Europe were linked to a kindred society," *Proceedings of the National Academy of Sciences of the United States of America* 116, no. 19 (May 7, 2019): 9471. Their identification codes are ans008 and ans014.

355 Ashkenazic J1c7a carriers in GenBank include samples HM627319, JF812166, and KY353087.

356 Costa, Pereira, et al., "A Substantial Prehistoric European Ancestry...," "Results" section.

357 Endre Neparáczki, Klaudia Kocsy, et al., "Revising mtDNA haplotypes of the ancient Hungarian conquerors with next generation sequencing," *PLoS ONE* 12, no. 4 (April 19, 2017): S1 Table. Its sample code is "Karos-III/4".

358 Stolarek, Juras, et al., "A mosaic genetic structure..." Her sample code is PCA0057.

359 Magdalena Fraser, Per Sjödin, et al., "The stone cist conundrum: A multidisciplinary approach to investigate Late Neolithic/Early Bronze Age population demography on the island of Gotland," *Journal of Archaeological Science: Reports* 20 (August 2018): Table 1.

360 Costa, Pereira, et al., "A Substantial Prehistoric European Ancestry...," Supplementary Note 2, p. 35.

361 Personal communication with Leo Cooper, March 25, 2021.

362 Pala, Olivieri, et al., "Mitochondrial DNA Signals of Late Glacial Recolonization of Europe from Near Eastern Refugia," Table S1 where their GenBank codes are listed as EU151466 and JQ797855 respectively.

363 These samples are coded as KC911442 and KC911605 in GenBank.

364 "Nadworna Shtetl Research Group - mtDNA Test Results for Members," Family Tree DNA, accessed January 13, 2022, https://www.familytreedna.com/public/NadwornaShtetl?iframe=mtresults

365 Pala, Olivieri, et al., "Mitochondrial DNA Signals of Late Glacial Recolonization of Europe from Near Eastern Refugia," supplementary data. The Lebanon sample is coded as JQ797965 and the Italy sample is coded as JQ797964 in GenBank.

366 "Cypriot_DNA - mtDNA Test Results for Members," Family Tree DNA, accessed January 13, 2022, https://www.familytreedna.com/public/Cypriot_DNA?iframe=mtresults

367 "Mountain Jewish DNA Project - mtDNA Test Results for Members," Family Tree DNA, accessed January 13, 2022, https://www.familytreedna.com/public/mountain-jewish-dna-project?iframe=mtresults

368 "Jewish Ukraine West - mtDNA Test Results for Members" where its K1a1b1 carrier names his matrilineal ancestor, Chava Goldberg from Poland.

369 Personal communication with Leo Cooper, May 4, 2021.

370 Maár, Varga, et al., "Maternal Lineages from 10–11th Century Commoner Cemeteries of the Carpathian Basin," Table S1 where she is identified as sample "Homokmégy-Székes/5".

371 Olalde, Mallick, et al., "The genomic history of the Iberian Peninsula over the past 8000 years," supplementary data. He was assigned the identifier I7643.

372 Haak, Lazaridis, et al., "Massive migration from the steppe...," Supplementary Data 1 where she is assigned the identifiers I0405 and Mina3.

373 Olalde, Brace, et al., "The Beaker phenomenon...," Supplementary Table 1. They were assigned the identifiers I0458 and I0461.

374 Olalde, Brace, et al., "The Beaker phenomenon...," Supplementary Table 1. They were assigned the identifiers I6587 and I6589.

375 Olalde, Mallick, et al., "The genomic history of the Iberian Peninsula over the past 8000 years," supplementary data. She was assigned the identifier I8141.

376 Olalde, Mallick, et al., "The genomic history of the Iberian Peninsula over the past 8000 years," supplementary data. She was assigned the identifier I12208.

377 Zalloua, Collins, et al., "Ancient DNA of Phoenician remains indicates discontinuity in the settlement history of Ibiza," Table 1. This sample's GenBank code is MH043585 and has the alternative identifier MS10589.

378 Rosa Fregel, Fernando L. Méndez, et al., "Ancient genomes from North Africa evidence prehistoric migrations to the Maghreb from both the Levant and Europe," *Proceedings of the National Academy of Sciences of the United States of America* 115, no. 26 (June 26, 2018): Tables S4.1 and S4.2 where these samples are listed as "KEB.3" and "KEB.4". Their GenBank codes are MF991436 and MF991437.

379 Olalde, Brace, et al., "The Beaker phenomenon...," Supplementary Table 1. He was assigned the identifier I6762.

380 Olalde, Brace, et al., "The Beaker phenomenon...," Supplementary Table 1. She was assigned the identifier I2651.

381 Olalde, Brace, et al., "The Beaker phenomenon...," Supplementary Table 1. She was assigned the identifier I7569.

382 Tina Saupe, Francesco Montinaro, et al., "Ancient genomes reveal structural shifts after the arrival of Steppe-related ancestry in the Italian peninsula," *Current Biology* 31, no. 9 (May 10, 2021): Table 1 where she has the identifier "BRC007/019".

383 Neus Font-Porterias, Neus Solé-Morata, et al., "The genetic landscape of Mediterranean North African populations through complete mtDNA sequences," *Annals of Human Biology* 45, no. 1 (February 2018): Supplementary Table 2 which codes them as BTUN01 and BTUN02. In GenBank their codes are MG182034 and MG182035.

384 Gerard Serra-Vidal, Marcel Lucas-Sanchez, et al., "Heterogeneity in Palaeolithic Population Continuity and Neolithic Expansion in North Africa," *Current Biology* 29, no. 22 (November 18, 2019): "Supplemental Information."

385 Claire-Elise Fischer, Anthony Lefort, et al., "The multiple maternal legacy of the Late Iron Age group of Urville-Nacqueville (France, Normandy) documents a long-standing genetic contact zone in northwestern France," *PLoS ONE* 13, no. 12 (December 6, 2018): S1 Table and S4 Table where this sample's number is 125.

386 Doron M. Behar, Ene Metspalu, et al., "The Matrilineal Ancestry of Ashkenazi Jewry: Portrait of a Recent Founder Event," *American Journal of Human Genetics* 78, no. 3 (March 2006): Table 3 and supplementary data. This sample's codes are D1606 in the study and DQ301798 in GenBank.

387 Costa, Pereira, et al., "A Substantial Prehistoric European Ancestry...," Supplementary Table S6.

388 Paweł Maciejko, *The Mixed Multitude: Jacob Frank and the Frankist Movement, 1755-1816* (Philadelphia: University of Pennsylvania Press, 2011), 1, 129, 195; Magda Teter, "Jewish Conversions to Catholicism in the Polish-Lithuanian Commonwealth of the Seventeenth and Eighteenth Centuries," *Jewish History* 17, no. 3 (2003): 258, 262. Apart from the Frankist mass conversions, Teter cited Jesuit records about Ashkenazic women who became Catholics, some of whom married Christian men.

389 Tomasz Grzybowski, Boris A. Malyarchuk, et al., "Complex interactions of the Eastern and Western Slavic populations with other European groups as revealed by mitochondrial DNA analysis," *Forensic Science International: Genetics* 1, no. 2 (June 2007): 144.

390 Grzybowski, Malyarchuk, et al., "Complex interactions...," 141, 142, 145.

391 For example, one Turkish Jew and six Bulgarian Jews with this haplogroup were included in: Behar, Metspalu, et al., "The Matrilineal Ancestry of Ashkenazi Jewry," Table 4. Similar carriers were found later in direct-to-consumer DNA tests.

392 Ferragut, Ramon, et al., "Middle Eastern genetic legacy in the paternal and maternal gene pools of Chuetas."

393 Gyaneshwer Chaubey, Manvendra Singh, et al., "Genetic affinities of the Jewish populations of India," *Scientific Reports* 6 (January 13, 2016): Supplementary Table 5.

394 Costa, Pereira, et al., "A Substantial Prehistoric European Ancestry...," "Results" section ("west European source") and "Introduction" section ("most likely assimilated in west (perhaps Mediterranean) Europe").

395 Chaubey, Singh, et al., "Genetic affinities of the Jewish populations of India," 6.

396 Shahzad Bhatti, Muhammad Aslamkhan, et al., "Genetic analysis of mitochondrial DNA control region variations in four tribes of Khyber Pakhtunkhwa, Pakistan," *Mitochondrial DNA Part A* 28, no. 5 (2017): "Abstract" section.

397 Zaman Stanizai, "Are Pashtuns the Lost Tribe of Israel?" Cambridge Open Engage, June 28, 2021, accessed January 31, 2022, https://www.cambridge.org/engage/coe/article-details/60d49c27c62295e4ef1ade46

398 Personal communication with Anthony Lizza, April 23, 2021.

399 Modi, Lancioni, et al., "The mitogenome portrait of Umbria in Central Italy as depicted by contemporary inhabitants and pre-Roman remains." The Umbrian sample is coded as MN687270 in GenBank.

400 Yves Gleize, Fanny Mendisco, et al., "Early Medieval Muslim Graves in France: First Archaeological, Anthropological and Palaeogenomic Evidence," *PLoS ONE* 11, no. 2 (February 24, 2016). He has the identifier SP7089.

401 Antonio, Gao, et al., "Ancient Rome," Table S2 where it she identified as sample R40.

402 Personal communication with Leo Cooper, May 30, 2021.

403 Olalde, Brace, et al., "The Beaker phenomenon...," Supplementary Table 1. She was assigned the identifier I2418.

404 Furtwängler, Rohrlach, et al., "Ancient genomes reveal social and genetic structure of Late Neolithic Switzerland." Their GenBank sample codes are MT079061, MT079062, MT079099, and MT079101.

405 Modi, Lancioni, et al., "The mitogenome portrait of Umbria in Central Italy as depicted by contemporary inhabitants and pre-Roman remains." This sample is coded as "aUMB044" in the study's data set and MN687310 in GenBank.

406 Ferragut, Ramon, et al., "Middle Eastern genetic legacy in the paternal and maternal gene pools of Chuetas," Table 3.

407 Costa, Pereira, et al., "A Substantial Prehistoric European Ancestry...," "Results" section.

408 Behar, Metspalu, et al., "The Matrilineal Ancestry of Ashkenazi Jewry," Table 4.

409 Mathieson, Lazaridis, et al., "Genome-wide patterns of selection in 230 ancient Eurasians," supplemental data. He has the identifier I0724.

410 Mathieson, Lazaridis, et al., "Genome-wide patterns of selection in 230 ancient Eurasians," supplemental data. He has the identifier I0634.

411 Olalde, Brace, et al., "The Beaker phenomenon...," Supplementary Table 1. He has the identifier I2630.

412 Iosif Lazaridis, Dani Nadel, et al., "Genomic insights into the origin of farming in the ancient Near East," *Nature* 536, no. 7617 (July 25, 2016). This sample has the identifier I0867.

413 Costa, Pereira, et al., "A Substantial Prehistoric European Ancestry...," "Introduction," "Results," and "Discussion" sections.

414 Shirin Farjadian, Marco Sazzini, et al., "Discordant Patterns of mtDNA and Ethno-Linguistic Variation in 14 Iranian Ethnic Groups," *Human Heredity* 72, no. 2 (2011): 82 (regarding K1a9 being in Iran in general) and supplementary data.

415 "Updated results of a Hungarian/Romanian and Iraqi/Syrian Turkmen," accessed January 12, 2022, https://www.reddit.com/r/23andme/comments/km180q/updated_results_of_a_hungarianromanian_and/

416 Behar, Metspalu, et al., "The Matrilineal Ancestry of Ashkenazi Jewry," Table 4.

417 Patterson, Isakov, et al., "Large-scale migration into Britain...," Supplementary Table 1 listing the relevant samples I14351 and I16403.

418 This sample is coded as MH120750 in GenBank.

419 Costa, Pereira, et al., "A Substantial Prehistoric European Ancestry...," "Results" section.

420 Ibid., Supplementary Note 1, p. 33.

421 Ibid., "Introduction" section.

422 Behar, Metspalu, et al., "The Matrilineal Ancestry of Ashkenazi Jewry," Table 4.

423 Personal communication with Leo Cooper, May 31, 2021.

424 F. A. Al-Jasmi's team sampled 21 Emiratis carrying K2a2a2. Their GenBank codes include MF437089, MF437100, MF437200, MF437238, and MF437256, among others. Additional Emiratis, plus the Saudi, tested with YFull.

425 van Straten, *Ashkenazic Jews and the Biblical Israelites*, 78.

426 Behar, Metspalu, et al., "The Matrilineal Ancestry of Ashkenazi Jewry," Table 4.

427 "L2a1 MtDNA--Africa and Beyond - mtDNA Test Results for Members," Family Tree DNA, accessed January 20, 2022, https://www.familytreedna.com/public/L2a1-mtdna?iframe=mtresults

428 Mielnik-Sikorska, Daca, et al., "The History of Slavs Inferred from Complete Mitochondrial Genome Sequences," "Haplogroup L Subclades in mtDNA Pools of Slavic Populations" section. This study's Polish L2a1l2a sample is coded as JX266264 in GenBank.

429 "L2a1 MtDNA--Africa and Beyond - mtDNA Test Results for Members."

430 Doron M. Behar, Richard Villems, et al., "The dawn of human matrilineal diversity," *American Journal of Human Genetics* 82, no. 5 (May 2008). This person is coded as L355 in their research and EU092721 in GenBank.

431 Candela L. Hernández, Pedro Soares, et al., "Early Holocenic and Historic mtDNA African Signatures in the Iberian Peninsula: The Andalusian Region as a Paradigm," *PLoS ONE* 10, no. 10 (October 28, 2015): S6 Dataset. These samples are coded as HG02620, HG02629, and HG02716.

432 This sample is coded as GMFUL5306377 in MTree.

433 Marina Silva, Farida Alshamali, et al., "60,000 years of interactions between Central and Eastern Africa documented by major African mitochondrial haplogroup L2," *Scientific Reports* 5 (July 27, 2015): Supplementary Table 3 where these samples are listed as members of this haplogroup that carry the GenBank codes JQ044978 and JQ044994.

434 González-Fores, Tassi, et al., "A western route of prehistoric human migration from Africa into the Iberian Peninsula." This sample is coded as MK321329 in GenBank.

435 Amy L. Non, Ali Al-Meeri, et al., "Mitochondrial DNA reveals distinct evolutionary histories for Jewish populations in Yemen and Ethiopia," *American Journal of Physical Anthropology* 144, no. 1 (January 2011): 3.

436 Costa, Pereira, et al., "A Substantial Prehistoric European Ancestry...," "Results" section.

437 Maár, Varga, et al., "Maternal Lineages from 10–11th Century Commoner Cemeteries of the Carpathian Basin," Table S1 where she is identified as sample 418 from that cemetery.

438 Jiao-Yang Tian, Hua-Wei Wang, et al., "A Genetic Contribution from the Far East into Ashkenazi Jews via the Ancient Silk Road," *Scientific Reports* 5 (February 11, 2015). The Sichuanese is coded as DJY576 on Table 1 and Figure 2 and as KP313562 in GenBank. EU148486 and KP313552 are Ashkenazic M33c carriers with matrilines from Belarus in GenBank.

439 The relevant sample is coded as KC990660 in GenBank.

440 This sample's GenBank code is FJ770968.

441 This sample's GenBank code is KP345999.

442 The Indian sample's GenBank code is JQ446400.

443 Inês Nogueiro, João C. Teixeira, et al., "Echoes from Sepharad: signatures on the maternal gene pool of crypto-Jewish descendants," *European Journal of Human Genetics* 23, no. 5 (May 2015): "Results and Discussion" section.

444 Agranat-Tamir, Waldman, et al., "The Genomic History of the Bronze Age Southern Levant," supplemental data. The Tel Megiddo sample has the identifier I10092, the Yehud sample is I7003, and the sample from Jordan is I3985.

445 Skourtanioti, Erdal, et al., "Genomic History of Neolithic to Bronze Age Anatolia, Northern Levant, and Southern Caucasus." This sample has the identifier ART032 in Table S9.

446 This sample's GenBank code is MN176236.

447 This sample's GenBank code is MG646144.

448 Emery, Duggan, et al., "Ancient Roman mitochondrial genomes...," supplementary data. The relevant sample has the identifiers "LRV 110" and F206.

449 "Jewish Ukraine West - mtDNA Test Results for Members."

450 Personal communication with Leo Cooper, May 30 and June 9, 2021.

451 Stefania Vai, Andrea Brunelli, et al., "A genetic perspective on Longobard-Era migrations," *European Journal of Human Genetics* 27, no. 4 (April 2019): supplementary data. One of these samples is coded as MG182464 in GenBank and alternatively known as SZ13. The other sample is SZ22.

452 Vai, Brunelli, et al., "A genetic perspective on Longobard-Era migrations," supplementary data. This sample is coded as MG182449 in GenBank and alternatively known as LRCMUS69.

453 Furtwängler, Rohrlach, et al., "Ancient genomes reveal social and genetic structure of Late Neolithic Switzerland," supplementary data. The relevant samples are coded as MT079106 and MT079091 in GenBank.

454 Agranat-Tamir, Waldman, et al., "The Genomic History of the Bronze Age Southern Levant," supplemental data including Table S1. It has the sample IDs I3832 and Levant39.

455 Costa, Pereira, et al., "A Substantial Prehistoric European Ancestry...," Supplementary Note 3, p. 41.

456 This person belongs to "Nadworna Shtetl Research Group - mtDNA Test Results for Members," which lists their matrilineal ancestor as Yocheved Haber, born in Nadworna, Galicia in 1848.

457 Deven N. Vyas, Andrew Kitchen, et al., "Bayesian analyses of Yemeni mitochondrial genomes suggest multiple migration events with Africa and Western Eurasia," *American Journal of Physical Anthropology* 159, no. 3 (March 2016): supplemental data. The sample from Yemen is coded as KM986527 in GenBank.

458 "The Haplogroup N mtDNA Study - mtDNA Test Results for Members," Family Tree DNA, accessed January 13, 2022, https://www.familytreedna.com/public/nmtdna/default.aspx?section=mtresults

459 Personal communication with Leo Cooper, March 27, 2021.

460 Several of Family Tree DNA's non-Jewish N9a3 carriers participate in public projects. Khamchiev has a matriline from Ingushetia, states "Orstkhoy-Bulguch nyaqan" (Orstkhoy were Chechen speakers and Bulgush is an Ingush name), and is in "Ingush DNA Project - mtDNA Test Results for Members," Family Tree DNA, accessed January 13, 2022, https://www.familytreedna.com/public/ingush?iframe=mtresults Another person with a matriline from Ingushetia (stating "Targimkhoy-Pliev," which are both Ingush surnames) as well as a person with a matriline from Chechnya (stating "Galay") and a person with a matriline from Moo Nyok Yin who was born circa 1900 in southeastern China are among the N9a3 carriers listed in "The Haplogroup N mtDNA Study - mtDNA Test Results for Members," Family Tree DNA, accessed January 13, 2022, https://www.familytreedna.com/public/nmtdna/default.aspx?section=mtresults

461 Some Chinese and Taiwanese carriers of N9a3 are coded as JQ705768, KC252517, KC252567, KU521464, and MT954929 in GenBank.

462 Miroslava Derenko, Boris A. Malyarchuk, et al., "Complete Mitochondrial DNA Analysis of Eastern Eurasian Haplogroups Rarely Found in Populations of Northern Asia and Eastern Europe," *PLoS ONE* 7, no. 2 (February 21, 2012).

463 The Buryat's GenBank code is JN857023.

464 Xiaoming Zhang, Chunmei Li, et al., "A Matrilineal Genetic Perspective of Hanging Coffin Custom in Southern China and Northern Thailand," *iScience* 23 (April 24, 2020): 3. This sample is coded as MN006852 in GenBank.

465 Derenko, Malyarchuk, et al., "Complete Mitochondrial DNA Analysis of Eastern Eurasian Haplogroups..." This Russian is identified as Rus_BGII-19 in this study and JN857057 in GenBank, while the Czech is Cz_V-44 and JN857038 respectively.

466 Péter L. Nagy, Judit Olasz, et al., "Determination of the phylogenetic origins of the Árpád Dynasty based on Y chromosome sequencing of Béla the Third," *European Journal of Human Genetics* 29, no. 1 (January 2021). These Bashkirs have the codes SRS6892124 and SRS6892238 in the BioProject database.

467 Personal communication with Leo Cooper, June 20, 2021.

468 Eszter Bánffy, Guido Brandt, et al., "'Early Neolithic' graves of the Carpathian Basin are in fact 6000 years younger—Appeal for real interdisciplinarity between archaeology and ancient DNA research," *Journal of Human Genetics* 57 (June 7, 2012): 467-468. The Sarmatian samples have the codes "Szakmár-Kisülés 8" and "Csongrád-Bokros 20" while the Hungarian sample is "Szarvas 23/20".

469 Choongwon Jeong, Ke Wang, et al., "A Dynamic 6,000-Year Genetic History of Eurasia's Eastern Steppe," *Cell* 183, no. 4 (November 12, 2020) where this sample is coded as IMA008.

470 Personal communication with Corey Bregman, June 11 and August 27, 2021.

471 Brook, "Sephardic Jews in Belarus," 6.

472 Ferragut, Ramon, et al., "Middle Eastern genetic legacy in the paternal and maternal gene pools of Chuetas."

473 Behar, Metspalu, et al., "Counting the Founders," "Results" section and Table 2; Costa, Pereira, et al., "A Substantial Prehistoric European Ancestry...," "Results" section.

474 Francesca Gandini, Alessandro Achilli, et al., "Mapping human dispersals into the Horn of Africa from Arabian Ice Age refugia using mitogenomes," *Scientific Reports* 6 (May 5, 2016): "Results" section under the subheading "An Arabian source for the major R0a lineages."

475 Lazaridis, Nadel, et al., "Genomic insights into the origin of farming in the ancient Near East." This sample has the identifiers I1699 and AG84_5.

476 Lazaridis, Nadel, et al., "Genomic insights into the origin of farming in the ancient Near East." This sample has the identifiers I1707 and AG83_5.

477 Harney, May, et al., "Ancient DNA from Chalcolithic Israel reveals the role of population mixture in cultural transformation," Supplementary Data 1 where this sample is given the identifier I1183.

478 "Jewish Ukraine West - mtDNA Test Results for Members."

479 Gandini, Achilli, et al., "Mapping human dispersals into the Horn of Africa...," "Results" section under the subheading "Deep ancestry of R0a."

480 Personal communication with Leo Cooper, May 31, 2021.

481 The French sample is coded as KF451056 in GenBank and HGDP00521 in the Human Genome Diversity Project.

482 Lazaridis, Nadel, et al., "Genomic insights into the origin of farming in the ancient Near East." This sample has the identifiers I1704 and AG89_1.

483 Relevant samples are coded as MK059509 and MK059544 in GenBank and LP166 and LP238 respectively in Jennifer Klunk's research.

484 Relevant samples are coded as MK059618 and MK059727 in GenBank and LE03 and LM73 respectively in Jennifer Klunk's research.

485 Endre Neparáczki, Zoltán Maróti, et al., "Mitogenomic data indicate admixture components of Central-Inner Asian and Srubnaya origin in the conquering Hungarians," *PLoS ONE* 13, no. 10 (October 18, 2018): supplementary data where the samples have the identifiers "Karos2/13", "Karos2/18", "Karos2/23", "Karos2/26", "Karos 2/32", and "Kenezlo-Fazekaszug2/1031"; Maár, Varga, et al., "Maternal Lineages from 10–11th Century Commoner Cemeteries of the Carpathian Basin," Table S1 where they are identified as samples 9, 16, and 151 from the Magyarhomorog-Kónyadomb cemetery.

486 Csáky, Gerber, et al., "Early medieval genetic data from Ural region...," especially Supplementary Information: Table S16 and Figure S4g.

487 Vai, Brunelli, et al., "A genetic perspective on Longobard-Era migrations," supplementary data. The relevant sample is coded as MG182461 in GenBank.

488 Haak, Lazaridis, et al., "Massive migration from the steppe...," Supplementary Data 1 where it is assigned the identifiers I0550 and KAR22A.

489 Olalde, Brace, et al., "The Beaker phenomenon...," Supplementary Table 1. They have the identifiers I5660 (RISE925) and I6624 (RISE932).

490   Olalde, Brace, et al., "The Beaker phenomenon...," Supplementary Table 1. She has the identifier I4892.

491   Olalde, Brace, et al., "The Beaker phenomenon...," Supplementary Table 1. He has the identifier I7628.

492   Anna Linderholm, Gülşah M. Kılınç, et al., "Corded Ware cultural complexity uncovered using genomic and isotopic analysis from south-eastern Poland," *Scientific Reports* 10 (April 14, 2020): Table 1 where he has the identifier "pcw250".

493   Mathieson, Alpaslan-Roodenberg, et al., "The genomic history of southeastern Europe," Supplementary Information where she is identified as sample I2105 (Yamna4).

494   This sample is coded as MK059514 in GenBank and has the alternative code LP173.

495   Neparáczki, Kocsy, et al., "Revising mtDNA haplotypes...," S1 Table where it is labeled "Karos-III/14".

496   Veronika Csáky, Dániel Gerber, et al., "Genetic insights into the social organisation of the Avar period elite in the 7th century AD Carpathian Basin," *Scientific Reports* 10 (January 22, 2020). This Avar was given the sample name HC9 in the study and MH894768 in GenBank.

497   Mittnik, Wang, et al., "The genetic prehistory of the Baltic Sea region," Table 1.

498   Behar, van Oven, et al., "A 'Copernican' reassessment of the human mitochondrial DNA tree from its root," supplementary data. The relevant sample is coded as JQ702925 in GenBank.

499   Personal communication with Leo Cooper, May 31, 2021.

500   Krzewińska, Kjellström, et al., "Genomic and Strontium Isotope Variation Reveal Immigration Patterns in a Viking Age Town," Table 1 on p. 2732 where he has the identifier "nuf002".

501   "Jewish Ukraine West - mtDNA Test Results for Members."

502   Personal communication with Leo Cooper, May 31, 2021.

503   This sample is coded as MK059659 in GenBank and also has the alternative identifier LG08.

504   Skourtanioti, Erdal, et al., "Genomic History of Neolithic to Bronze Age Anatolia, Northern Levant, and Southern Caucasus." This sample has the identifier ART042 in Table S9.

505   Costa, Pereira, et al., "A Substantial Prehistoric European Ancestry...," "Results" section.

506   Pertinent samples from Turkey, Iran, and Russia are included in Pala, Olivieri, et al., "Mitochondrial DNA Signals of Late Glacial Recolonization of Europe from Near Eastern Refugia," Table S2.

507   Costa, Pereira, et al., "A Substantial Prehistoric European Ancestry...," "Results" section.

508   Personal communication with Leo Cooper, May 31, 2021.

509   Sandra Wilde, Adrian Timpson, et al., "Direct evidence for positive selection of skin, hair, and eye pigmentation in Europeans during the last 5,000 y.," *Proceedings of the National Academy of Sciences of the United States of America* 111, no. 13 (April 1, 2014). The sample carries the identifier code POP1.

510   Haak, Lazaridis, et al., "Massive migration from the steppe...," Supplementary Data 1 where it is assigned the identifiers I0106 and ESP26.

511   Emery, Duggan, et al., "Ancient Roman mitochondrial genomes...," supplementary data where it has the identifier "LRV 137". It is coded as MG773644 in GenBank.

512 Piotrowska-Nowak, Elson, et al., "New mtDNA Association Model…," supplementary data. It is coded.as MH120744 in GenBank.

513 Fregel, Méndez, et al., "Ancient genomes from North Africa…," Table S4.2 and p. 32 of "Supplementary Notes" where this sample is listed as "KEB.7". Its GenBank code is MF991439.

514 Fregel, Méndez, et al., "Ancient genomes from North Africa…," Table S4.2 and p. 32 of "Supplementary Notes" where these samples are listed as "TOR.6" and "TOR.7". Their GenBank codes are MF991443 and MF991444.

515 Diroma, Modi, et al., "New Insights Into Mitochondrial DNA Reconstruction and Variant Detection in Ancient Samples," Figure 2. The relevant sample is coded as MW389256 in GenBank.

516 Matisoo-Smith, Gosling, et al., "Ancient mitogenomes of Phoenicians from Sardinia and Lebanon," Table 1. The relevant sample is coded as KY797250 in GenBank.

517 Matisoo-Smith, Gosling, et al., "Ancient mitogenomes of Phoenicians from Sardinia and Lebanon," "Discussion" section.

518 Olivieri, Sidore, et al., "Mitogenome Diversity in Sardinians." This sample is coded as KY399146 in GenBank.

519 Silva, Oteo-García, et al., "Biomolecular insights…," supplemental data. These samples are coded as MZ920763 and MZ921073 in GenBank.

520 Maár, Varga, et al., "Maternal Lineages from 10–11th Century Commoner Cemeteries of the Carpathian Basin," Table S1 where it is identified as sample "Nagytarcsa-Homokbánya/20".

521 Dariusz Błaszczyk, "The Early Middle Ages," in *Early Medieval and Early Modern Burial Site in Pień*, ed. Dariusz Poliński (Toruń, Poland: Wydawnictwo Edukacyjne "Akapit", 2020), 269-270.

522 The relevant sample is coded as MK059550 in GenBank and LP250 in Jennifer Klunk's research.

523 "T2b Mitochondria Haplogroup Project - mtDNA Test Results for Members," Family Tree DNA, accessed January 13, 2022, https://www.familytreedna.com/public/T2b/default.aspx?section=mtresults

524 Personal communication with Leo Cooper, May 31, 2021.

525 Krishna R. Veeramah, Andreas Rott, et al., "Population genomic analysis of elongated skulls reveals extensive female-biased immigration in Early Medieval Bavaria," *Proceedings of the National Academy of Sciences of the United States of America* 115, no. 13 (March 27, 2018): supplementary data where this sample is coded as BIM_37.

526 Stolarek, Juras, et al., "A mosaic genetic structure…" They assigned this sample the identifier PCA0026.

527 Frasera, Sjödin, et al., "The stone cist conundrum," Table 1 on p. 327 where this sample's code is "hgb002".

528 Personal communication with Leo Cooper, June 8, 2021.

529 A T2b25 carrier lists the matrilineal ancestor Anna Diamantaras from Greece in "T2b Mitochondria Haplogroup Project - mtDNA Test Results for Members."

530 Personal communication with Anthony Lizza, April 19, 2021.

531 Malyarchuk, Litvinov, et al., "Mitogenomic diversity in Russians and Poles," supplemental data. The relevant sample is coded as KY670852 in GenBank.

532 Silva, Oteo-García, et al., "Biomolecular insights…," supplemental data. These samples are coded as MZ920261, MZ920977, and MZ921011 in GenBank.

533 Malyarchuk, Derenko, et al., "Whole mitochondrial genome diversity in two Hungarian populations," "Abstract" section.

534 Felice L. Bedford, Doron Yacobi, et al., "Clarifying Mitochondrial DNA Subclades of T2e from Mideast to Mexico," *Journal of Phylogenetics and Evolutionary Biology* 1, no. 4 (2013): 4 and supplementary data. In GenBank, the Sephardic lineage from Romania is coded as KF564289 and the Sephardic lineage from the Netherlands is coded as KF048033.

535 Nogueiro, Teixeira, et al., "Echoes from Sepharad," "Results and Discussion" section; Nogueiro, Teixeira, et al., "Portuguese crypto-Jews," 8.

536 Vai, Brunelli, et al., "A genetic perspective on Longobard-Era migrations," supplementary data. The relevant sample is coded as MG182498 in GenBank.

537 Verónica Fernandes, Petr Triska, et al., "Genetic Stratigraphy of Key Demographic Events in Arabia," *PLoS ONE* 10, no. 3 (March 4, 2015): "Supplementary Materials: S2 Table." The relevant sample is coded as JN086657 in GenBank.

538 Personal communication with Leo Cooper, March 27, 2021.

539 This sample is coded as MF437096 in GenBank.

540 Personal communication with Anthony Lizza, April 19, 2021.

541 This sample is coded as MK059652 in GenBank and LE17 in Jennifer Klunk's research.

542 Harney, May, et al., "Ancient DNA from Chalcolithic Israel reveals the role of population mixture in cultural transformation," Supplementary Data 1 where this sample is given the identifier I1154.

543 "Mountain Jewish DNA Project - mtDNA Test Results for Members."

544 Behar, Metspalu, et al., "Counting the Founders," Table 2.

545 Fernandes, Triska, et al., "Genetic Stratigraphy of Key Demographic Events in Arabia," "Supplementary Materials: S2 Table." This sample carries the code EF556186 in GenBank.

546 The Iraqi sample is coded as JQ705079 and the Italian sample is JQ798127 in GenBank.

547 Israel Pickholtz, "The Woman Who Matches Etta Bryna," April 6, 2014, http://allmyforeparents.blogspot.com/2014/04/the-woman-who-matches-etta-bryna.html and personal communication with him, October 25, 2020.

548 An Armenian U1b1 sample from Turkey is coded as MK491447 in GenBank.

549 The relevant sample is coded as both KF451170 and KJ445908 in GenBank and HGDP00647 in the Human Genome Diversity Project.

550 Two relevant samples are coded as KY409272 and KY409361 in GenBank.

551 Antonio, Gao, et al., "Ancient Rome," Table S2 where she is identified as R114 and "burial 114".

552 Nogueiro, Teixeira, et al., "Portuguese crypto-Jews," "mtDNA in Sephardic Portuguese Jews" section.

553 Mathieson, Lazaridis, et al., "Genome-wide patterns of selection in 230 ancient Eurasians," supplemental data. He has the identifiers I1541 and ESP32.

554 Olalde, Brace, et al., "The Beaker phenomenon...," Supplementary Table 1. She has the identifiers I5023 and RISE565.

555 Joachim Burger, Vivian Link, et al., "Low Prevalence of Lactase Persistence in Bronze Age Europe Indicates Ongoing Strong Selection over the Last 3,000 Years," *Current Biology* 30, no. 21 (November 2, 2020): Table S1 where his identifier is WEZ15.

556 Anna Juras, Przemysław Makarowicz, et al., "Mitochondrial genomes from Bronze Age Poland reveal genetic continuity from the Late Neolithic and additional genetic affinities with the steppe populations," *American Journal of Physical Anthropology* 172, no. 2 (June 2020) where this sample is assigned the identifier "poz541". This sample is in GenBank under the code MN699890.

557 Anna Juras, Maciej Chyleński, et al., "Mitochondrial genomes reveal an east to west cline of steppe ancestry in Corded Ware populations," *Scientific Reports* 8 (August 2, 2018): Table 1 where his sample is coded as "poz090".

558 Narasimhan, Patterson, et al., "The formation of human populations in South and Central Asia," Table S1 and pp. 33-34 of Supplementary Materials where she has the identifiers I0244 and SVP62.

559 It was already in North Yorkshire circa 2400-1500 B.C.E. per Patterson, Isakov, et al., "Large-scale migration into Britain...," Supplementary Table 1 listing the relevant sample I16400.

560 These samples are coded as MK321335, MT079093, MT079095, and MK059535 in GenBank.

561 The German's GenBank sample code is MW182431 and it is listed as Dorothea's lineage at the public family tree, Geni, accessed September 16, 2021, https://www.geni.com/people/Dorothea/6000000114766619100

562 Stolarek, Juras, et al., "A mosaic genetic structure...," supplementary data where they are identified with the sample codes PCA0028 and PCA0054.

563 Ireneusz Stolarek, Luiza Handschuh, et al., "Goth migration induced changes in the matrilineal genetic structure of the central-east European population," *Scientific Reports* 9 (May 1, 2019): Table 1 on p. 4 where they are identified with the sample codes PCA0088 and PCA0090.

564 "Jewish Ukraine West - mtDNA Test Results for Members."

565 Jennifer Klunk, Ana T. Duggan, et al., "Genetic resiliency and the Black Death: No apparent loss of mitogenomic diversity due to the Black Death in medieval London and Denmark," *American Journal of Physical Anthropology* 169, no. 2 (June 2019). The relevant sample is coded as LM60 in this study and MK059716 in GenBank.

566 Patterson, Isakov, et al., "Large-scale migration into Britain...," Supplementary Table 1 listing the relevant sample I20985 dated to circa 450-1 B.C.E.

567 Miroslava Derenko, Galina Denisova, et al., "Insights into matrilineal genetic structure, differentiation and ancestry of Armenians based on complete mitogenome data," *Molecular Genetics and Genomics* 294, no. 6 (December 2019). The relevant sample is coded as MK491389 in GenBank.

568 Alexander S. Semenov and Vladimir V. Bulat, "Ancient Paleo-DNA of Pre-Copper Age North-Eastern Europe," *European Journal of Molecular Biotechnology* 11, no. 1 (2016): Table 1.

569 Mathieson, Alpaslan-Roodenberg, et al., "The genomic history of southeastern Europe."

570 Neparáczki, Maróti, et al., "Mitogenomic data indicate admixture components...," supplementary data where the sample has the identifier "Karos3/16". In GenBank its code is KY083708.

571 Haak, Lazaridis, et al., "Massive migration from the steppe...," Supplementary Data 1 where she is assigned the identifiers I0171 and BZH12.

572 Olalde, Brace, et al., "The Beaker phenomenon...," Supplementary Table 1. This sample was given the identifier I2421.

573 Allentoft, Sikora, et al., "Population genomics of Bronze Age Eurasia," supplemental data where this sample is coded as RISE496.
574 Boris A. Malyarchuk, Miroslava Derenko, et al., "The peopling of Europe from the mitochondrial haplogroup U5 perspective," *PLoS ONE* 5, no. 4 (April 21, 2010). These samples are coded as GU296604 and GU296616 in GenBank.
575 Jing Zhao, Wurigemule Wurigemule, et al., "Genetic substructure and admixture of Mongolians and Kazakhs inferred from genome-wide array genotyping," *Annals of Human Biology* 47, no. 7-8 (December 2020).
576 The relevant sample is coded as LD49 in Jennifer Klunk's research.
577 The relevant sample is coded as MK059611 in GenBank and LD91 in Jennifer Klunk's research.
578 Vai, Brunelli, et al., "A genetic perspective on Longobard-Era migrations," supplementary data. The relevant sample is coded as MG182459 in GenBank and alternatively known as LHHEG60.
579 Anna Juras, Maja Krzewińska, et al., "Diverse origin of mitochondrial lineages in Iron Age Black Sea Scythians," *Scientific Reports* 7 (March 7, 2017): "Results" section. This sample is coded as KX977302 in GenBank.
580 Mathieson, Alpaslan-Roodenberg, et al., "The genomic history of southeastern Europe," Supplementary Information where they are identified as samples I3718, I5879, I5881, and I5890.
581 Ashot Margaryan, Miroslava Derenko, et al., "Eight Millennia of Matrilineal Genetic Continuity in the South Caucasus," *Current Biology* 27, no. 13 (July 10, 2017): Deposited Data. The relevant sample is coded as MF362706 in GenBank and has the alternative identifier "arm24".
582 Eppie R. Jones, Gunita Zarina, et al., "The Neolithic Transition in the Baltic Was Not Driven by Admixture with Early European Farmers," *Current Biology* 27, no. 4 (February 20, 2017): supplemental data. This sample has the identifiers I4629 and ZVEJ28.
583 Juras, Chyleński, et al., "Mitochondrial genomes reveal...," Table 1 where his sample is coded as "poz232".
584 Vai, Brunelli, et al., "A genetic perspective on Longobard-Era migrations," supplementary data. The relevant sample is coded as MG182533 in GenBank.
585 Malyarchuk, Derenko, et al., "The peopling of Europe from the mitochondrial haplogroup U5 perspective." This sample is coded as GU296628 in GenBank.
586 Maár, Varga, et al., "Maternal Lineages from 10–11th Century Commoner Cemeteries of the Carpathian Basin," Table S1 where it is identified as sample "Vörs-Papkert-B/434".
587 Jukka Kiiskilä, Jukka S. Moilanen, et al., "Analysis of functional variants in mitochondrial DNA of Finnish athletes," *BMC Genomics* 20 (October 29, 2019): supplemental data. This Finnish sample is coded as MN516614 in GenBank.
588 Malyarchuk, Litvinov, et al., "Mitogenomic diversity in Russians and Poles," supplemental data. This sample comes from Russia's Oryol region and is coded as KY670955 in GenBank.
589 Malyarchuk, Derenko, et al., "Mitogenomic diversity in Tatars from the Volga-Ural region of Russia." This Tatar's sample is coded as GU123032 in GenBank.
590 Nagy, Olasz, et al., "Determination of the phylogenetic origins of the Árpád Dynasty based on Y chromosome sequencing of Béla the Third." This Bashkir has the code SRS6892285 in the BioProject database. YFull places this person into U5a1d2b1.

591 Peng, Xu, et al., "Mitochondrial genomes uncover the maternal history of the Pamir populations," supplementary data. This Pamiri's sample is coded as MF523107 in GenBank.

592 Rem I. Sukernik, Natalia V. Volodko, et al., "Mitochondrial genome diversity in the Tubalar, Even, and Ulchi: contribution to prehistory of native Siberians and their affinities to Native Americans," *American Journal of Physical Anthropology* 148, no. 1 (May 2012): supplementary data. This study's Tubular sample is coded as FJ147317 in GenBank and an unpublished study's Tubular sample is MG660604.

593 These Uyghur samples are coded as KU683149 and KU683359 in GenBank.

594 Miroslava Derenko, Boris A. Malyarchuk, et al., "Complete Mitochondrial DNA Diversity in Iranians," *PLoS ONE* 8, no. 11 (November 14, 2013): supplementary data. The Persian's GenBank code is KC911432.

595 Maár, Varga, et al., "Maternal Lineages from 10–11th Century Commoner Cemeteries of the Carpathian Basin," Table S1 where they are identified as samples "Homokmégy-Székes/215", "Vörs-Papkert-B/51", and "Nagytarcsa-Homokbánya/1".

596 Martina Unterländer, Friso Palstra, et al., "Ancestry and demography and descendants of Iron Age nomads of the Eurasian Steppe," *Nature Communications* 8 (March 3, 2017): supplementary data where this sample has the identifier Pr13.

597 Unterländer, Palstra, et al., "Ancestry and demography...," supplementary data where this sample has the identifier A_9.

598 Damgaard, Marchi, et al., "137 ancient human genomes from across the Eurasian steppes," supplementary data. She was assigned the identifiers DA50 and "Kyr 18".

599 Mittnik, Wang, et al., "The genetic prehistory of the Baltic Sea region," Table 1. This sample is coded as Tamula1 in the study and MG429018 in GenBank.

600 Allentoft, Sikora, et al., "Population genomics of Bronze Age Eurasia," supplemental data where this sample is coded as RISE546.

601 Narasimhan, Patterson, et al., "The formation of human populations in South and Central Asia," supplemental data. These samples have been assigned the identifiers I3388, I3949, and I6714.

602 Järve, Saag, et al., "Shifts in the Genetic Landscape...," Tables S1 and S2 where this sample is coded as MJ-51.

603 "Jewish Ukraine West - mtDNA Test Results for Members."

604 The Poland sample is coded as GU296587 and the Slovakia sample as GU296646 in GenBank.

605 Personal communication with Leo Cooper, May 31, 2021.

606 Vai, Brunelli, et al., "A genetic perspective on Longobard-Era migrations," supplementary data. The relevant sample is coded as MG182455 in GenBank.

607 This sample is coded as MG429007 in GenBank.

608 Juras, Makarowicz, et al., "Mitochondrial genomes from Bronze Age Poland..." where these samples are assigned the identifiers "poz155" and "poz549".

609 Chaubey, Singh, et al., "Genetic affinities of the Jewish populations of India," Supplementary Table 5.

610 Polish examples are coded as MF177151, MF177152, MH120765, MH120810, MH120684, MN176234 in GenBank.

611 "U5b FMS mtDNA Project - mtDNA Test Results for Members," Family Tree DNA, accessed January 22, 2022, https://www.familytreedna.com/public/U5b_FGS?iframe=mtresults

612 The samples are coded as HGDP00694 and HGDP00695 in the Human Genome Diversity Project.

613 Juras, Makarowicz, et al., "Mitochondrial genomes from Bronze Age Poland..." where this sample is assigned the identifier "poz650".

614 Margaryan, Lawson, et al., "Population genomics of the Viking world." They are identified as samples VK51 and VK251 in this study.

615 Margaryan, Lawson, et al., "Population genomics of the Viking world." She is identified as sample VK196 in this study.

616 This sample is coded as MK059616 in GenBank and LD98 in Jennifer Klunk's research.

617 Personal communication with Leo Cooper, May 31, 2021.

618 "Albanian U5b2a1a," April 6, 2013, accessed March 19, 2022, https://www.eupedia.com/forum/threads/28514-Albanian-U5b2a1a

619 Malyarchuk, Derenko, et al., "The peopling of Europe from the mitochondrial haplogroup U5 perspective," Table S3.

620 Anna Juras, Maciej Chyleński, et al., "Investigating kinship of Neolithic post-LBK human remains from Krusza Zamkowa, Poland using ancient DNA," *Forensic Science International: Genetics* 26 (2017): Table 1 on p. 35 where her sample code is KZ1.

621 Mathieson, Alpaslan-Roodenberg, et al., "The genomic history of southeastern Europe," Supplementary Information where they are identified as samples I4871 and I4874.

622 Mathieson, Alpaslan-Roodenberg, et al., "The genomic history of southeastern Europe," Supplementary Information where he is identified as sample I4550.

623 Mathieson, Alpaslan-Roodenberg, et al., "The genomic history of southeastern Europe," Supplementary Information where they are identified as samples I3716 and I6133.

624 Mathieson, Alpaslan-Roodenberg, et al., "The genomic history of southeastern Europe," Supplementary Information where she is identified as sample I2519.

625 Olalde, Brace, et al., "The Beaker phenomenon...," Supplementary Table 1. She was assigned the identifier I0456.

626 Olalde, Brace, et al., "The Beaker phenomenon...," Supplementary Table 1. They have the identifiers I7579 and I7580.

627 Maciej Chyleński, Edvard Ehler, et al., "Ancient Mitochondrial Genomes Reveal the Absence of Maternal Kinship in the Burials of Çatalhöyük People and Their Genetic Affinities," *Genes (Basel)* 10, no. 3 (March 2019): Table 1. This sample's GenBank code is MK308701.

628 Sécher, Fregel, et al., "The history of the North African mitochondrial DNA haplogroup U6 gene flow into the African, Eurasian and American continents," Supplement 1. Among the samples they used, the Algerian has the GenBank code JX120727, the Mexican has JX120752, the Spaniard has JX120751, and the Italian has EF064337.

629 One of these samples from Spain has the GenBank code KT819220 and comes from this study: Hernández, Soares, et al., "Early Holocenic and Historic mtDNA African Signatures in the Iberian Peninsula," "Results" section. Another U6a7a1b carrier participates in "U6 haplogroup mitochondrial DNA project - mtDNA Test Results for Members," Family Tree DNA, accessed January 13, 2022, https://www.

familytreedna.com/public/U6mtdna?iframe=mtresults and lists the matrilineal ancestor Josefa Garcia Labella, born in 1900 in Spain.

630 David W. von Ehrlicher, "mt-DNA haplogroup U6a7a1b in Zacatecas and Jessica Alba," February 2, 2021, accessed April 21, 2021, in Nuestros Ranchos, http://www.nuestrosranchos.com/en/node/24829

631 "U6 haplogroup mitochondrial DNA project - mtDNA Test Results for Members," where a U6a7a1b matriline from Portugal lists the ancestor Maria Pinto who was born in the 1650s.

632 This sample has the code KT779171 in GenBank.

633 Rosa Fregel, Alejandra C. Ordóñez, et al., "Mitogenomes illuminate the origin and migration patterns of the indigenous people of the Canary Islands," *PLoS ONE* 14, no. 3 (March 20, 2019).

634 Joseph H. Marcus, Cosimo Posth, et al., "Genetic history from the Middle Neolithic to present on the Mediterranean island of Sardinia," *Nature Communications* 11 (February 24, 2020): supplementary data where she is coded as AMC001.

635 Mariekevan de Loosdrecht, Abdeljalil Bouzouggar, et al., "Pleistocene North African genomes link Near Eastern and sub-Saharan African human populations," *Science* 360, no. 6388 (May 4, 2018): Table S13 where they are listed with the codes TAF011 and TAF012.

636 Fregel, Méndez, et al., "Ancient genomes from North Africa...," Tables S4.1 and S4.2 where this sample is listed as "IAM.6".

637 "mtDNA Haplogroup Mutations," Family Tree DNA, accessed January 12, 2022, https://www.familytreedna.com/mtdna-haplogroup-mutations.aspx

638 Hovhannes Sahakyan, Baharak H. Kashani, et al., "Origin and spread of human mitochondrial DNA haplogroup U7," *Scientific Reports* 7 (April 7, 2017): Supplementary Dataset 1, Table S1. The Turkish sample is coded as KY824883 in GenBank. The Armenian sample from Artsakh is MF362816.

639 Costa, Pereira, et al., "A Substantial Prehistoric European Ancestry...," Supplementary Note 3, p. 42.

640 Matisoo-Smith, Gosling, et al., "Ancient mitogenomes of Phoenicians from Sardinia and Lebanon," supplemental data. This sample is coded as KY797190 in GenBank.

641 This sample is coded as KY824908 in GenBank.

642 Peidong Shen, Tal Lavi, et al., "Reconstruction of Patrilineages and Matrilineages of Samaritans and Other Israeli Populations From Y-Chromosome and Mitochondrial DNA Sequence Variation," *Human Mutation* 24, no. 3 (September 2004): 255.

643 Relevant samples are coded as MF362929 and MF362935 in GenBank.

644 Silva, Oteo-García, et al., "Biomolecular insights...," supplemental data. In GenBank, these samples are MZ920284 from Valencia and MZ920603 from Galicia.

645 Modi, Lancioni, et al., "The mitogenome portrait of Umbria in Central Italy as depicted by contemporary inhabitants and pre-Roman remains." The ancient Umbri sample is coded as MN687315 in GenBank.

646 Diroma, Modi, et al., "New Insights Into Mitochondrial DNA Reconstruction and Variant Detection in Ancient Samples," Figure 2. The relevant samples are coded as MW389258, MW389268, and MW389273 in GenBank.

647 Daniel M. Fernandes, Alissa Mittnik, et al., "The spread of steppe and Iranian-related ancestry in the islands of the western Mediterranean," *Nature Ecology and Evolution* 4, no. 3 (March 2020): supplementary data where this sample is identified as I7805.

648   Fernandes, Mittnik, et al., "The spread of steppe and Iranian-related ancestry in the islands of the western Mediterranean," supplementary data where this sample is identified as I3123.

649   Antonio, Gao, et al., "Ancient Rome," Table S2 where she is identified as sample R24.

650   Mathieson, Alpaslan-Roodenberg, et al., "The genomic history of southeastern Europe," Supplementary Information where he is identified as sample I3498.

651   Olalde, Brace, et al., "The Beaker phenomenon...," Supplementary Table 1. They were assigned the identifiers I7214 and I7289.

652   Olalde, Brace, et al., "The Beaker phenomenon...," Supplementary Table 1. This sample was given the identifier I4894.

653   Furtwängler, Rohrlach, et al., "Ancient genomes reveal social and genetic structure of Late Neolithic Switzerland." This sample is coded as MT079105 in GenBank.

654   Maár, Varga, et al., "Maternal Lineages from 10–11th Century Commoner Cemeteries of the Carpathian Basin," Table S1 where it is identified as sample "Homokmégy-Székes/92".

655   Mathieson, Lazaridis, et al., "Genome-wide patterns of selection in 230 ancient Eurasians," supplemental data. He has the identifiers I0745 and M11-363.

656   Schuenemann, Peltzer, et al., "Ancient Egyptian mummy genomes suggest an increase of Sub-Saharan African ancestry in post-Roman periods," supplemental data where this sample carries the identifier JK2951.

657   This sample is coded as MK059558 in GenBank and LP385 in Jennifer Klunk's research.

658   Margaryan, Lawson, et al., "Population genomics of the Viking world." He is identified as sample VK144 in this study and also has the alternate identifier "UK_Oxford_#8".

659   Juras, Makarowicz, et al., "Mitochondrial genomes from Bronze Age Poland..." where this sample is assigned the identifier "poz584" and the GenBank code MN699904.

660   Ioana Rusu, Alessandra Modi, et al., "Maternal DNA lineages at the gate of Europe in the 10th century AD," *PLoS ONE* 13, no. 3 (March 14, 2020): "Results and Discussion" section. This sample has been assigned the code MF597774 in GenBank.

661   These samples are coded as MF498721 and MK059490 in GenBank respectively.

662   Behar, van Oven, et al., "A 'Copernican' reassessment of the human mitochondrial DNA tree from its root," supplementary data. This sample is coded as JQ703830 in GenBank.

663   Li, Besenbacher, et al., "Variation and association to diabetes in 2000 full mtDNA sequences mined from an exome study in a Danish population." This sample is coded as KF162774 in GenBank.

664   Piotrowska-Nowak, Kosior-Jarecka, et al., "Investigation of whole mitochondrial genome variation in normal tension glaucoma." These Polish samples are coded as MG646181, MG646234, and MG646238 in GenBank.

665   Malyarchuk, Litvinov, et al., "Mitogenomic diversity in Russians and Poles," supplemental data. The relevant sample is coded as KY670883 in GenBank.

666   Maár, Varga, et al., "Maternal Lineages from 10–11th Century Commoner Cemeteries of the Carpathian Basin," Table S1 where they are identified as samples "Homokmégy-Székes/241", "Homokmégy-Székes/41", "Ibrány Esbóhalom/161", "Magyarhomorog-Kónyadomb/2", and "Sárrétudvari-Hízóföld/110".

667   Personal communication with Leo Cooper, May 31, 2021.

668   "Lituania Propria - mtDNA Test Results for Members," Family Tree DNA, accessed June 13, 2021, https://www.familytreedna.com/public/lituaniapropria/default.aspx?section=mtresults

669   Margaryan, Lawson, et al., "Population genomics of the Viking world." He is identified as sample VK362 in this study.

670   "V mtDNA - Haplogroup V - mtDNA Test Results for Members," Family Tree DNA, accessed January 9, 2022, https://www.familytreedna.com/public/mtdnahaplogroupv?iframe=mtresults which includes a V15 carrier descending matrilineally from Poppa de Bayeux in 9th-century France.

671   This sample is coded as KM047196 in GenBank.

672   Silva, Oteo-García, et al., "Biomolecular insights...," supplemental data. This sample is coded as MZ921003 in GenBank.

673   "V mtDNA - Haplogroup V - mtDNA Test Results for Members."

674   This sample is coded as MG773623 in GenBank, has the library identifier "LIAV 7" in Matthew Emery's research, and has the Italian identifier "Scavi Latanzi F1, 5".

675   Raule, Sevini, et al., "The co-occurrence of mtDNA mutations...," supplementary data. This sample is coded as JX153072 in GenBank.

676   Olivieri, Pala, et al., "Mitogenomes from Two Uncommon Haplogroups...," supplemental data. This sample is coded as KF146271 in GenBank.

677   These samples are coded as MT588229, MT588254, and MT588259 in GenBank.

678   "Jewish Ukraine West - mtDNA Test Results for Members."

679   Personal communication with Leo Cooper, May 31, 2021.

680   Malyarchuk, Derenko, et al., "Mitogenomic diversity in Tatars from the Volga-Ural region of Russia."

681   Maár, Varga, et al., "Maternal Lineages from 10–11th Century Commoner Cemeteries of the Carpathian Basin," Table S1 where he is identified as sample 197 from that cemetery.

682   Haak, Lazaridis, et al., "Massive migration from the steppe...," Extended Data Table 2.

683   Juras, Chyleński, et al., "Mitochondrial genomes reveal...," Table 1.

684   Olalde, Brace, et al., "The Beaker phenomenon...," Supplementary Table 1.

685   Gnecchi-Ruscone, Khussainova, et al., "Ancient genomic time transect from the Central Asian Steppe unravels the history of the Scythians," "Supplementary Materials" and supplementary data where these samples have the identifiers "NUL001.A", "BRE011.A", and "KNT003.A".

686   A person with an Ashkenazic matriline from Lithuania is in GenBank as sample MT551785.

687   Personal communication with Leo Cooper, June 8, 2021.

688   Margaryan, Derenko, et al., "Eight Millennia of Matrilineal Genetic Continuity in the South Caucasus," Deposited Data. The relevant sample is coded as MF362715 in GenBank.

689   "Jewish Ukraine West - mtDNA Test Results for Members."

690   Public family tree of Jeanne (Beaudinet) Bajolet, WikiTree, accessed September 13, 2021, https://www.wikitree.com/wiki/Beaudinet-1

691   "mtDNA Test Results for Members," Acadian Heritage DNA project, Family Tree DNA, accessed September 16, 2021, https://www.familytreedna.com/public/acadianheritage?iframe=mtresults under the section listing X2b7-carrying matrilineal descendants of Beaudinet's daughter Barbe Bajolet.

692 García, Alonso, et al., "Forensically relevant phylogeographic evaluation of mitogenome variation in the Basque Country." The relevant samples are coded as MN046428 and MN046459 in GenBank.

693 Silva, Oteo-García, et al., "Biomolecular insights...," supplemental data. This sample is coded as MZ920581 in GenBank.

694 Ivan G. Marcus, "Why Did Medieval Northern French Jewry (Şarfat) Disappear?" in *Jews, Christians and Muslims in Medieval and Early Modern Times*, ed. Arnold E. Franklin et al. (Leiden: Brill, 2014), 99-117.

695 Derenko, Malyarchuk, et al., "Western Eurasian ancestry in modern Siberians based on mitogenomic data."

696 Liran I. Shlush, Doron M. Behar, "The Druze: a population genetic refugium of the Near East," *PLoS ONE* 3, no. 5 (May 7, 2008).

697 Emery, Duggan, et al., "Ancient Roman mitochondrial genomes...," supplementary data. The relevant sample is coded as MG773647 in GenBank and also carries the identifiers LRV 93 and F96a.

698 Skourtanioti, Erdal, et al., "Genomic History of Neolithic to Bronze Age Anatolia, Northern Levant, and Southern Caucasus." This sample has the identifier ALA004 in Table S9.

699 Derenko, Malyarchuk, et al., "Western Eurasian ancestry in modern Siberians based on mitogenomic data."

700 Verónica Fernandes, Farida Alshamali, et al., "The Arabian Cradle: Mitochondrial Relicts of the First Steps along the Southern Route out of Africa," *American Journal of Human Genetics* 90, no. 2 (February 10, 2012).

701 The Armenian samples are coded as MF362765, MF362901, and MF362943 in GenBank.

702 Behar, Metspalu, et al., "Counting the Founders."

703 Felice L. Bedford, "Sephardic signature in haplogroup T mitochondrial DNA," *European Journal of Human Genetics* 20, no. 4 (April 2012).

704 "Romaniote DNA Project - mtDNA Test Results for Members."

705 Luísa Pereira, Nuno M. Silva, et al., "Population expansion in the North African late Pleistocene signalled by mitochondrial DNA haplogroup U6," *BMC Evolutionary Biology* 10 (December 21, 2010): supplemental data. This Bulgarian Jew's GenBank code is HQ651714.

706 Pereira, Silva, et al., "Population expansion in the North African late Pleistocene...," supplemental data. This Bulgarian Jew's GenBank code is HQ651713.

707 Behar, Metspalu, et al., "Counting the Founders," Table 2.

708 Ibid.

709 Ibid.

710 Personal communication with Leo Cooper, May 10, 2021.

711 Brook, *The Jews of Khazaria*, 244-245.

712 Behar, Metspalu, et al., "Counting the Founders," Table 2; Costa, Pereira, et al., "A Substantial Prehistoric European Ancestry...," Supplementary Table S7; Chaubey, Singh, et al., "Genetic affinities of the Jewish populations of India," Supplementary Table 5 where the Bene Israel Jews are called "Mumbai Jewish".

713 Brook, *The Jews of Khazaria*, 208. In the nineteenth century, Krymchaks spoke a Turkic language based on Crimean Tatar.

714 Personal communication with Leo Cooper, June 22, 2021.

715 Mikhail Kizilov, *The Karaites of Galicia: An Ethnoreligious Minority among the Ashkenazim, the Turks, and the Slavs, 1772-1945* (Leiden: Brill, 2009), 215-216, 219-221. Prominent rabbis banned this intermarriage.

716 Ibid., 318 indicates that this community is nearly extinct.

717 Mikhail Kizilov, *The Sons of Scripture: The Karaites in Poland and Lithuania in the Twentieth Century* (Berlin: De Gruyter Open, 2015).

718 Many Karaites migrated from the cities of Istanbul and Edirne to the Crimea, per Dan Shapira, "Beginnings of the Karaites of the Crimea Prior to the Early Sixteenth Century," in *A Guide to Karaite Studies: An Introduction to the Literary Sources of Medieval and Modern Karaite Judaism*, ed. Meira Polliack (Leiden: Brill, 2003), 725.

719 Brook, *The Jews of Khazaria*, 213-214; Brook, "The Genetics of Crimean Karaites," 73-76, 80, 83. For Anastasiya T. Agdzhoyan's 2018 dissertation "Genofond korennykh narodov Kryma po markeram Y-khromosomy, mtDNK i polnogenomnykh paneley autosomnykh SNP," she sampled the Y-DNA haplogroup C3 from several people identifying as Crimean Karaites. I believe it probably results from recent intermarriage with Crimean Tatars because that population has C3 and it is of Mongol origin. To date, Dubrovitzky and I have been unable to verify in objective and reliable published sources that any old, bona fide Karaite family lines have C3 or any other Turkic or East Eurasian haplogroup, even though Karaites in eastern Europe spoke Turkic languages for centuries.

720 Brook, *The Jews of Khazaria*, 115; Brook, "The Genetics of Crimean Karaites," 78-79.

721 This sample is coded as K16 in Brook, "The Genetics of Crimean Karaites," 79 and is listed in the public table of "Karaites of E.Europe - mtDNA Test Results for Members."

722 Personal communication with Leon Kull, January 13, 2022.

723 Personal communication with Ian Logan, January 12, 2022.

724 One Turkey H2a1i carrier is sample SRS8752599 in the BioProject database and another is ERS1789540 in YFull.

725 A relevant sample is coded as KC911386 in GenBank.

726 A relevant sample is coded as JQ703760 in GenBank.

727 Ferragut, Ramon, et al., "Middle Eastern genetic legacy in the paternal and maternal gene pools of Chuetas," Table 3.

728 Ibid., "Haplogroup composition" in "Results and Discussion" section.

729 Nogueiro, Teixeira, et al., "Echoes from Sepharad."

730 Nogueiro, Teixeira, et al., "Portuguese crypto-Jews," "mtDNA in Sephardic Portuguese Jews" section.

731 Ibid., Figure 4 on p. 9.

732 Some Ashkenazim in Poland maintained Jewish and Polish identities simultaneously. For example, Arthur Rubinstein, *My Young Years* (New York: Alfred A. Knopf, 1973), 10: "We spoke Polish at home, I was a Pole. ...I loved Poland. In the afternoon I had Polish lessons with my sister Frania which were a source of great pleasure to me." Rubinstein, p. 46, also wrote that his family had pride in having Jewish ancestry even though they were not religiously observant. Other Polish-speaking Jews identifying as Poles practiced Judaism per Paweł Jasnowski, "The Failure of the Integration of Galician Jews According to Lvov's *Ojczyzna* (1881-1892)," *Scripta Judaica Cracoviensia* 13 (2015): 57-61, 64.

733 Hundert, *Jews in Poland-Lithuania in the Eighteenth Century*, 77.

734 Delphine Bechtel, "Cultural Transfers between 'Ostjuden' and 'Westjuden': German-Jewish Intellectuals and Yiddish Culture 1897–1930," *The Leo Baeck Institute Year Book* 42, no. 1 (January 1997): 67–68.

735 Tobias Grill, "Preface," in *Jews and Germans in Eastern Europe: Shared and Comparative Histories*, ed. Tobias Grill (Berlin: De Gruyter, 2018), XI–XII.

736 Kateřina Čapková, "Czechs, Germans, Jews — Where is the Difference? The Complexity of National Identities of Bohemian Jews, 1918-1938," *Bohemia* 46, no. 1 (2005): 13.

737 G. A. McDowell, E. H. Mules, et al., "The presence of two different infantile Tay-Sachs disease mutations in a Cajun population," *American Journal of Human Genetics* 51, no. 5 (November 1992).

738 Mazal Karpati, Leah Peleg, et al., "A novel mutation in the HEXA gene specific to Tay-Sachs disease carriers of Jewish Iraqi origin," *Clinical Genetics* 57, no. 5 (May 2000).

739 Their impression was largely built from the autosomal analysis in Gil Atzmon, Li Hao, et al., "Abraham's Children in the Genome Era: Major Jewish Diaspora Populations Comprise Distinct Genetic Clusters with Shared Middle Eastern Ancestry," *American Journal of Human Genetics* 86, no. 6 (June 11, 2010): 854 which stated that "Of the European populations, the Northern Italians showed the greatest proximity to the Jews, followed by Sardinians and French . . ." It was amplified by Jon Entine's headline "Ashkenazi Jewish women descended mostly from Italian converts, new study asserts," Genetic Literacy Project, October 8, 2013, accessed January 19, 2022, https://geneticliteracyproject.org/2013/10/08/ashkenazi-jewish-women-descended-mostly-from-italian-converts-new-study-asserts/ which discussed Costa's team's subsequent mtDNA study. Steven M. Bray, Jennifer G. Mulle, et al. cautioned in "Signatures of founder effects, admixture, and selection in the Ashkenazi Jewish population," *Proceedings of the National Academy of Sciences of the United States of America* 107, no. 37 (September 14, 2010): 16223–16224 that "the proximity of the AJ and Italian populations . . . is also consistent that the projection of the AJ populations is primarily the outcome of admixture with Central and Eastern European hosts that coincidentally shift them closer to Italians along principle component axes relative to Middle Easterners." Identical-by-descent analysis in subsequent years did not support the idea that Northern Italians were the most important European source for Ashkenazic ancestry.

740 Akihiro Fujimoto, Ryosuke Kimura, et al., "A Scan for Genetic Determinants of Human Hair Morphology: EDAR Is Associated with Asian Hair Thickness," *Human Molecular Genetics* 17, no. 6 (March 15, 2008): 835–843.

741 Brook, *The Jews of Khazaria*, 39–41.

742 Ibid., chap. 6 and appendix D.

743 Igor V. Kornienko, Tatiana G. Faleeva, et al., "Y-Chromosome Haplogroup Diversity in Khazar Burials from Southern Russia," *Russian Journal of Genetics* 57, no. 4 (April 2021): 477–488. This article discusses Y-DNA and autosomal DNA results.

744 Alexander S. Mikheyev, Lijun Qiu, et al., "Diverse genetic origins of medieval steppe nomad conquerors," bioRxiv, December 16, 2019, accessed October 3, 2021, https://www.biorxiv.org/content/10.1101/2019.12.15.876912v1

745 Gennadii Afanasyev, Sh. Vien', et al., "Khazarskie konfederaty v Basseyne Dona," in *Yestestvennonauchnie metodi issledovaniya i paradigma sovremennoy arkheologii:*

Materiali Vserossiyskoy nauchnoy konferentsii, Moskva, Institut arkheologii Rossiyskoy akademii nauk, 8–11 dekabrya 2015 (Moscow: Yaziki slavyanskoy kul'tury, 2015), 146–153.

746 Mari Réthelyi, "Hungarian Jewish Stories of Origin: Samuel Kohn, the Khazar Connection and the Conquest of Hungary," *Hungarian Cultural Studies: e-Journal of the American Hungarian Educators Association* 14 (2021): 54–61.

747 Doron M. Behar, Mait Metspalu, et al., "No Evidence from Genome-Wide Data of a Khazar Origin for the Ashkenazi Jews," *Human Biology* 85, no. 6 (December 2013): 882.

748 Ibid., 883.

749 Cooper, post #66, https://anthrogenica.com/showthread.php?23847-How-much-Middle-Eastern-ancestry-do-Ashkenazim-have/page7

750 Dzhaubermezov, Gabidullina, et al., "Ancient DNA analysis of Early Medieval Alan populations of the North Caucasus," Figure 4.

751 On the earlier Mizrahi Jewish origin for Ashkenazic R2a, see Brook, *The Jews of Khazaria*, 185.

752 Alexander Beider, *A Dictionary of Ashkenazic Given Names: Their Origins, Structure, Pronunciations, and Migrations* (Bergenfield, NJ: Avotaynu, 2001), 197.

Bibliography

Achilli, Alessandro, Chiara Rengo, et al. "The Molecular Dissection of mtDNA Haplogroup H Confirms That the Franco-Cantabrian Glacial Refuge Was a Major Source for the European Gene Pool." *American Journal of Human Genetics* 75, no. 5 (December 2004): 910–918.

Afanasyev, Gennadii, Sh. Vien', et al. "Khazarskie konfederaty v Basseyne Dona." In *Yestestvennonauchnie metodi issledovaniya i paradigma sovremennoy arkheologii: Materiali Vserossiyskoy nauchnoy konferentsii, Moskva, Institut arkheologii Rossiyskoy akademii nauk, 8–11 dekabrya 2015*, 146–153. Moscow: Yaziki slavyanskoy kul'tury, 2015.

Agranat-Tamir, Lily, Shamam Waldman, et al. "The Genomic History of the Bronze Age Southern Levant." *Cell* 181, no. 5 (May 28, 2020): 1146–1157.e11. doi:10.1016/j.cell.2020.04.024.

Albrechtsen, Anders, N. Grarup, et al., "Exome sequencing-driven discovery of coding polymorphisms associated with common metabolic phenotypes," *Diabetologia* 56, no. 2 (2013): 298–310.

Allen, Wayne. *The Cantor: From the Mishnah to Modernity*. Eugene, OR: Wipf and Stock, 2019.

Allentoft, Morten E., Martin Sikora, et al. "Population genomics of Bronze Age Eurasia." *Nature* 522, no. 7555 (June 11, 2015): 167–172.

Antonio, Margaret L., Ziyue Gao, et al. "Ancient Rome: A genetic crossroads of Europe and the Mediterranean." *Science* 366, no. 6466 (November 8, 2019): 708–714.

Atzmon, Gil, Li Hao, et al. "Abraham's Children in the Genome Era: Major Jewish Diaspora Populations Comprise Distinct Genetic Clusters with Shared Middle Eastern Ancestry." *American Journal of Human Genetics* 86, no. 6 (June 11, 2010): 850–859.

Avraham, Alexander. "Sephardim." In *The YIVO Encyclopedia of Jews in Eastern Europe*, edited by Gershon D. Hundert, vol. 2, 1689–1692. New Haven, CT: Yale University Press, 2008.

Bánffy, Eszter, Guido Brandt, et al. "'Early Neolithic' graves of the Carpathian Basin are in fact 6000 years younger—Appeal for real interdisciplinarity between archaeology and ancient DNA research." *Journal of Human Genetics* 57 (June 7, 2012): 467–469.

Bechtel, Delphine. "Cultural Transfers between 'Ostjuden' and 'Westjuden': German-Jewish Intellectuals and Yiddish Culture 1897–1930." *The Leo Baeck Institute Year Book* 42, no. 1 (January 1997): 67–83.

Bedford, Felice L. "Sephardic signature in haplogroup T mitochondrial DNA." *European Journal of Human Genetics* 20, no. 4 (April 2012): 441–448.

Bedford, Felice L., Doron Yacobi, et al. "Clarifying Mitochondrial DNA Subclades of T2e from Mideast to Mexico." *Journal of Phylogenetics and Evolutionary Biology* 1, no. 4 (2013): 1–8.

Behar, Doron M., Bayazit Yunusbayev, et al. "The Genome-Wide Structure of the Jewish People." *Nature* 466, no. 7303 (July 8, 2010): 238–242.

Behar, Doron M., Christine Harmant, et al. "The Basque Paradigm: Genetic Evidence of a Maternal Continuity in the Franco-Cantabrian Region since Pre-Neolithic Times." *American Journal of Human Genetics* 90, no. 3 (March 9, 2012): 486–493.

Behar, Doron M., Ene Metspalu, et al. "The Matrilineal Ancestry of Ashkenazi Jewry: Portrait of a Recent Founder Event." *American Journal of Human Genetics* 78, no. 3 (March 2006): 487–497.

Behar, Doron M., Ene Metspalu, et al. "Counting the Founders: The Matrilineal Genetic Ancestry of the Jewish Diaspora." *PLoS ONE* 3, no. 4 (April 30, 2008): e2062. doi:10.1371/journal.pone.0002062.

Behar, Doron M., Lauri Saag, et al. "The genetic variation in the R1a clade among the Ashkenazi Levites' Y chromosome." *Scientific Reports* 7 (November 2, 2017): article no. 14969. doi:10.1038/s41598-017-14761-7.

Behar, Doron M., Mait Metspalu, et al. "No Evidence from Genome-Wide Data of a Khazar Origin for the Ashkenazi Jews." *Human Biology* 85, no. 6 (December 2013): 859–900.

Behar, Doron M., Mannis van Oven, et al. "A 'Copernican' reassessment of the human mitochondrial DNA tree from its root." *American Journal of Human Genetics* 90, no. 4 (April 6, 2012): 675–684.

Behar, Doron M., Michael F. Hammer, et al. "MtDNA Evidence for a Genetic Bottleneck in the Early History of the Ashkenazi Jewish Population." *European Journal of Human Genetics* 12, no. 5 (May 2004): 355–364.

Behar, Doron M., Richard Villems, et al. "The dawn of human matrilineal diversity." *American Journal of Human Genetics* 82, no. 5 (May 2008): 1130–1140.

Beider, Alexander. *A Dictionary of Ashkenazic Given Names: Their Origins, Structure, Pronunciations, and Migrations.* Bergenfield, NJ: Avotaynu, 2001.

Beider, Alexander. "Names and Naming." In *The YIVO Encyclopedia of Jews in Eastern Europe*, edited by Gershon D. Hundert, vol. 1, 1248–1251. New Haven, CT: Yale University Press, 2008.

Beider, Alexander. *Origins of Yiddish Dialects.* Oxford: Oxford University Press, 2015.

Beider, Alexander. "Exceptional Ashkenazic Surnames of Sephardic Origin." *Avotaynu: The International Review of Jewish Genealogy* 33, no. 4 (Winter 2017): 3–5.

Bhatti, Shahzad, Muhammad Aslamkhan, et al. "Genetic analysis of mitochondrial DNA control region variations in four tribes of Khyber Pakhtunkhwa, Pakistan." *Mitochondrial DNA Part A* 28, no. 5 (2017): 687–697.

Blady, Ken. *Jewish Communities in Exotic Places*. Northvale, NJ: Jason Aronson, 2000.

Błaszczyk, Dariusz. "The Early Middle Ages." In *Early Medieval and Early Modern Burial Site in Pień*, edited by Dariusz Poliński, 257–275. Toruń, Poland: Wydawnictwo Edukacyjne "Akapit", 2020.

Bodner, Martin, Alessandra Iuvaro, et al. "Helena, the hidden beauty: Resolving the most common West Eurasian mtDNA control region haplotype by massively parallel sequencing an Italian population sample." *Forensic Science International: Genetics* 15 (March 2015): 21–26. doi:10.1016/j.fsigen.2014.09.012.

Bray, Steven M., Jennifer G. Mulle, et al. "Signatures of founder effects, admixture, and selection in the Ashkenazi Jewish population." *Proceedings of the National Academy of Sciences of the United States of America* 107, no. 37 (September 14, 2010): 16222–16227.

Brook, Kevin Alan. "The Origins of East European Jews." *Russian History/Histoire Russe* 30, no. 1–2 (Spring-Summer 2003): 1–22.

Brook, Kevin Alan. "The Genetics of Crimean Karaites." *Karadeniz Araştırmaları* no. 42 (Summer 2014): 69–84.

Brook, Kevin Alan. "Sephardic Jews in Galitzian Poland and Environs." *ZichronNote: The Journal of the San Francisco Bay Area Jewish Genealogical Society* 36, no. 2 (May 2016): 5–6, 11–12.

Brook, Kevin Alan. "Sephardic Jews in Lithuania and Latvia." *ZichronNote: The Journal of the San Francisco Bay Area Jewish Genealogical Society* 36, no. 3 (August 2016): 9–11.

Brook, Kevin Alan. "Sephardic Jews in Central and Northern Poland." *ZichronNote: The Journal of the San Francisco Bay Area Jewish Genealogical Society* 37, no. 1–2 (February/May 2017): 18–20.

Brook, Kevin Alan. "Sephardic Jews in Belarus." *ZichronNote: The Journal of the San Francisco Bay Area Jewish Genealogical Society* 38, no. 1–2 (February/May 2018): 5–6.

Brook, Kevin Alan. "Sephardic Jews in Central, Eastern, and Southern Ukraine." *ZichronNote: The Journal of the San Francisco Bay Area Jewish Genealogical Society* 39, no. 2 (May 2019): 14–15.

Brook, Kevin Alan. "More Evidence for Sephardic Jews in Poland." *ZichronNote: The Journal of the San Francisco Bay Area Jewish Genealogical Society* 41, no. 3 (August 2021): 13–14.

Brook, Kevin Alan. *The Jews of Khazaria*, 3rd ed. Lanham, MD: Rowman & Littlefield, 2018.

Burger, Joachim, Vivian Link, et al. "Low Prevalence of Lactase Persistence in Bronze Age Europe Indicates Ongoing Strong Selection over the Last 3,000 Years." *Current Biology* 30, no. 21 (November 2, 2020): 4307–4315.e13. doi:10.1016/j.cub.2020.08.033.

Buś, Magdalena M., Maria Lembring, et al. "Mitochondrial DNA analysis of a Viking age mass grave in Sweden." *Forensic Science International: Genetics* 42 (September 2019): 268–274. doi:10.1016/j.fsigen.2019.06.002.

Čapková, Kateřina. "Czechs, Germans, Jews — Where is the Difference? The Complexity of National Identities of Bohemian Jews, 1918–1938." *Bohemia* 46, no. 1 (2005): 7–14.

Cassidy, Lara M., Ros Ó Maoldúin, et al. "A dynastic elite in monumental Neolithic society." *Nature* 582, no. 7812 (June 17, 2020): 384–388.

Cassidy, Lara M., Rui Martiniano, et al. "Neolithic and Bronze Age migration to Ireland and establishment of the insular Atlantic genome." *Proceedings of the National Academic of Sciences of the United States of America* 113, no. 2 (January 12, 2016): 368–373.

Chaubey, Gyaneshwer, Manvendra Singh, et al. "Genetic affinities of the Jewish populations of India." *Scientific Reports* 6 (January 13, 2016): article no. 19166. doi:10.1038/srep19166.

Chazan, Robert. *The Jews of Medieval Western Christendom: 1000–1500*. Cambridge: Cambridge University Press, 2006.

Chyleński, Maciej, Edvard Ehler, et al. "Ancient Mitochondrial Genomes Reveal the Absence of Maternal Kinship in the Burials of Çatalhöyük People and Their Genetic Affinities." *Genes (Basel)* 10, no. 3 (March 2019): article no. 207. doi:10.3390/genes10030207.

Clemente, Florian, Martina Unterländer, et al. "The genomic history of the Aegean palatial civilizations." *Cell* 184, no. 10 (May 13, 2021): 2565–2586.e21. doi:10.1016/j.cell.2021.03.039.

Coffman-Levy, Ellen. "A mosaic of people: the Jewish story and a reassessment of the DNA evidence." *Journal of Genetic Genealogy* 1 (2005): 12–33. Accessed October 3, 2021. https://jogg.info/pages/11/coffman.pdf

Collins, David W., Harini V. Gudiseva, et al. "Association of primary open-angle glaucoma with mitochondrial variants and haplogroups common in African Americans." *Molecular Vision* 22 (May 16, 2016): 454–471. Accessed October 4, 2021. https://pubmed.ncbi.nlm.nih.gov/27217714/

Costa, Marta D., Joana B. Pereira, et al. "A Substantial Prehistoric European Ancestry amongst Ashkenazi Maternal Lineages." *Nature Communications* 4 (October 8, 2013): article no. 2543. doi:10.1038/ncomms3543.

Csáky, Veronika, Dániel Gerber, et al. "Genetic insights into the social organisation of the Avar period elite in the 7th century AD Carpathian Basin." *Scientific Reports* 10 (January 22, 2020): article no. 948. doi:10.1038/s41598–019–57378–8.

Csáky, Veronika, Dániel Gerber, et al. "Early medieval genetic data from Ural region evaluated in the light of archaeological evidence of ancient Hungarians." *Scientific Reports* 10 (November 5, 2020): article no. 19137. doi:10.1038/s41598–020–75910-z.

Damgaard, Peter de Barros, Nina Marchi, et al. "137 ancient human genomes from across the Eurasian steppes." *Nature* 557, no. 7705 (May 9, 2018): 369–374.

De Fanti, Sara, Chiara Barbieri, et al. "Fine Dissection of Human Mitochondrial DNA Haplogroup HV Lineages Reveals Paleolithic Signatures from European Glacial Refugia." *PLoS ONE* 10, no. 12 (December 7, 2015): e0144391. doi:10.1371/journal.pone.0144391.

Derenko, Miroslava, Boris A. Malyarchuk, et al. "Phylogeographic analysis of mitochondrial DNA in northern Asian populations." *American Journal of Human Genetics* 81, no. 5 (November 2007): 1025–1041.

Derenko, Miroslava, Boris A. Malyarchuk, et al. "Complete Mitochondrial DNA Analysis of Eastern Eurasian Haplogroups Rarely Found in Populations of Northern Asia and Eastern Europe." *PLoS ONE* 7, no. 2 (February 21, 2012): e32179. doi:10.1371/journal.pone.0032179.

Derenko, Miroslava, Boris A. Malyarchuk, et al. "Complete Mitochondrial DNA Diversity in Iranians." *PLoS ONE* 8, no. 11 (November 14, 2013): e80673. doi:10.1371/journal.pone.0080673.

Derenko, Miroslava, Boris A. Malyarchuk, et al. "Western Eurasian ancestry in modern Siberians based on mitogenomic data." *BMC Evolutionary Biology* 14 (October 10, 2014): article no. 217. doi:10.1186/s12862-014-0217-9.

Derenko, Miroslava, Galina Denisova, et al. "Mitogenomic diversity and differentiation of the Buryats." *Journal of Human Genetics* 63, no. 1 (January 2018): 71–81.

Derenko, Miroslava, Galina Denisova, et al. "Insights into matrilineal genetic structure, differentiation and ancestry of Armenians based on complete mitogenome data." *Molecular Genetics and Genomics* 294, no. 6 (December 2019): 1547–1559.

Diroma, Maria Angela, Alessandra Modi, et al. "New Insights Into Mitochondrial DNA Reconstruction and Variant Detection in Ancient Samples." *Frontiers in Genetics* 12 (February 18, 2021): 619950. doi:10.3389/fgene.2021.619950.

Dittmann, Robert. "West Slavic Canaanite Glosses in Medieval Hebrew Manuscripts." *Judaica: Beiträge zum Verstehen des Judentums* 73, no. 2–3 (June-September 2017): 234–283.

Drineas, Petros, Fotis Tsetsos, et al. "Genetic history of the population of Crete." *Annals of Human Genetics* (June 13, 2019): article no. 12328. doi:10.1111/ahg.12328.

Dzhaubermezov, Murat, Liliia Gabidullina, et al. "Ancient DNA analysis of Early Medieval Alan populations of the North Caucasus." Paper presented at Widening Horizons: 27th Annual Meeting of the European Association of Archaeologists, Kiel, Germany, September 8–11, 2021. Repository, session no. 412, paper and abstract no. 2378. Prague: European Association of Archaeologists, 2021.

Elliott, Katherine S., Marc Haber, et al. "Fine-scale genetic structure in the United Arab Emirates reflects endogamous and consanguineous culture, population history and geography." *Molecular Biology and Evolution* 39, no. 3 (March 2022): msac039. doi:10.1093/molbev/msac039.

Emery, Matthew V., Ana T. Duggan, et al. "Ancient Roman mitochondrial genomes and isotopes reveal relationships and geographic origins at the local and pan-Mediterranean scales." *Journal of Archaeological Science: Reports* 20 (August 2018): 200–209.

Engel, Edna. "Immigrant Scribes' Handwriting in Northern Italy from the Late Thirteenth to the Mid-Sixteenth Century: Sephardi and Ashkenazi Attitudes toward the Italian Script." In *The Late Medieval Hebrew Book in the Western Mediterranean: Hebrew Manuscripts and Incunabula in Context*, edited by Javier del Barco, 28–45. Leiden: Brill, 2015.

Entine, Jon. *Abraham's Children: Race, Identity, and the DNA of the Chosen People*. New York: Grand Central Publishing, 2007.

Farjadian, Shirin, Marco Sazzini, et al. "Discordant Patterns of mtDNA and Ethno-Linguistic Variation in 14 Iranian Ethnic Groups." *Human Heredity* 72, no. 2 (2011): 73–84.

Fernandes, Daniel M., Alissa Mittnik, et al. "The spread of steppe and Iranian-related ancestry in the islands of the western Mediterranean." *Nature Ecology and Evolution* 4, no. 3 (March 2020): 334–345.

Fernandes, Verónica, Farida Alshamali, et al. "The Arabian Cradle: Mitochondrial Relicts of the First Steps along the Southern Route out of Africa." *American Journal of Human Genetics* 90, no. 2 (February 10, 2012): 347–355.

Fernandes, Verónica, Petr Triska, et al. "Genetic Stratigraphy of Key Demographic Events in Arabia." *PLoS ONE* 10, no. 3 (March 4, 2015): e0118625. doi:10.1371/journal. pone.0118625.

Ferragut, Joana F., Cristian Ramon, et al. "Middle Eastern genetic legacy in the paternal and maternal gene pools of Chuetas." *Scientific Reports* 10 (December 8, 2020): article no. 21428. doi:10.1038/s41598-020-78487-9.

Figueras, Pau. "Epigraphic Evidence for Proselytism in Ancient Judaism." *Immanuel* 24/25 (1990): 194–206.

Fischer, Claire-Elise, Anthony Lefort, et al. "The multiple maternal legacy of the Late Iron Age group of Urville-Nacqueville (France, Normandy) documents a long-standing genetic contact zone in northwestern France." *PLoS ONE* 13, no. 12 (December 6, 2018): e0207459. doi:10.1371/journal.pone.0207459.

Fleming, K. E. *Greece—A Jewish History*. Princeton, NJ: Princeton University Press, 2008.

Font-Porterias, Neus, Neus Solé-Morata, et al. "The genetic landscape of Mediterranean North African populations through complete mtDNA sequences." *Annals of Human Biology* 45, no. 1 (February 2018): 98–104.

Fraser, Magdalena, Per Sjödin, et al. "The stone cist conundrum: A multidisciplinary approach to investigate Late Neolithic/Early Bronze Age population demography on the island of Gotland." *Journal of Archaeological Science: Reports* 20 (August 2018): 324–337.

Fregel, Rosa, Alejandra C. Ordóñez, et al. "Mitogenomes illuminate the origin and migration patterns of the indigenous people of the Canary Islands." *PLoS ONE* 14, no. 3 (March 20, 2019): e0209125. doi:10.1371/journal.pone.0209125.

Fregel, Rosa, Fernando L. Méndez, et al. "Ancient genomes from North Africa evidence prehistoric migrations to the Maghreb from both the Levant and Europe." *Proceedings of the National Academy of Sciences of the United States of America* 115, no. 26 (June 26, 2018): 6774–6779.

Fujimoto, Akihiro, Ryosuke Kimura, et al. "A Scan for Genetic Determinants of Human Hair Morphology: EDAR Is Associated with Asian Hair Thickness." *Human Molecular Genetics* 17, no. 6 (March 15, 2008): 835–843.

Furtwängler, Anja, Adam B. Rohrlach, et al. "Ancient genomes reveal social and genetic structure of Late Neolithic Switzerland." *Nature Communications* 11 (April 20, 2020): article no. 1915. doi:10.1038/s41467–020–15560–x.

Gamba, Cristina, Eppie R. Jones, et al. "Genome flux and stasis in a five millennium transect of European prehistory." *Nature Communications* 5 (October 21, 2014): article no. 5257. doi:10.1038/ncomms6257.

Gandini, Francesca, Alessandro Achilli, et al. "Mapping human dispersals into the Horn of Africa from Arabian Ice Age refugia using mitogenomes." *Scientific Reports* 6 (May 5, 2016): article no. 25472. doi:10.1038/srep25472.

García, Óscar, Santos Alonso, et al. "Forensically relevant phylogeographic evaluation of mitogenome variation in the Basque Country." *Forensic Science International: Genetics* 46 (May 2020): article no. 102260. doi:10.1016/j.fsigen.2020.102260.

García-Fernández, Carla, Neus Font-Porterias, et al. "Sex-biased patterns shaped the genetic history of Roma." *Scientific Reports* 10 (September 2, 2020): article no. 14464. doi:10.1038/s41598–020–71066–y.

Gleize, Yves, Fanny Mendisco, et al. "Early Medieval Muslim Graves in France: First Archaeological, Anthropological and Palaeogenomic Evidence." *PLoS ONE* 11, no. 2 (February 24, 2016): e0148583. doi:10.1371/journal.pone.0148583.

Glick, Leonard B. *Abraham's Heirs: Jews and Christians in Medieval Europe*. Syracuse, NY: Syracuse University Press, 1999.

Gnecchi-Ruscone, Guido Alberto, Elmira Khussainova, et al. "Ancient genomic time transect from the Central Asian Steppe unravels the history of the Scythians." *Science Advances* 7, no. 13 (March 26, 2021): eabe4414. doi:10.1126/sciadv.abe4414.

Goldberg, Jacob. "Tavernkeeping." In *The YIVO Encyclopedia of Jews in Eastern Europe*, edited by Gershon D. Hundert, vol. 2, 1849–1852. New Haven, CT: Yale University Press, 2008.

Golden, Peter B. "The Conversion of the Khazars to Judaism." In *The World of the Khazars: New Perspectives—Selected Papers from the Jerusalem 1999 International Khazar Colloquium*, edited by Peter B. Golden, Haggai Ben-Shammai, and András Róna-Tas, 123–162. Leiden: Brill, 2007.

Goldstein, David B. *Jacob's Legacy: A Genetic View of Jewish History*. New Haven, CT: Yale University Press, 2008.

González-Fores, Gloria, F. Tassi, et al. "A western route of prehistoric human migration from Africa into the Iberian Peninsula." *Proceedings of the Royal Society B: Biological*

Sciences 286, no. 1895 (January 23, 2019): article no. 20182288. doi:10.1098/rspb.2018.2288.

Grill, Tobias. "Preface." In *Jews and Germans in Eastern Europe: Shared and Comparative Histories*, edited by Tobias Grill, VII-XXII. Berlin: De Gruyter, 2018.

Grzybowski, Tomasz, Boris A. Malyarchuk, et al. "Complex interactions of the Eastern and Western Slavic populations with other European groups as revealed by mitochondrial DNA analysis." *Forensic Science International: Genetics* 1, no. 2 (June 2007): 141–147. doi:10.1016/j.fsigen.2007.01.010.

Gurianov, Vladimir, Dmitry Adamov, et al. "Clarification of Y-DNA Haplogroup Q1b Phylogenetic Structure Based on Y-Chromosome Full Sequencing." *Russian Journal of Genetic Genealogy* 7, no. 1 (March 31, 2015): 90–105. Accessed October 3, 2021. https://www.academia.edu/11738905/Clarification_of_Y-DNA_Haplogroup_Q1b_Phylogenetic_Structure_Based_on_Y-Chromosome_Full_Sequencing

Haak, Wolfgang, Iosif Lazaridis, et al. "Massive migration from the steppe was a source for Indo-European languages in Europe." *Nature* 522, no. 7555 (June 11, 2015): 207–211.

Haber, Marc, Joyce Nassar, et al. "A Genetic History of the Near East from an aDNA Time Course Sampling Eight Points in the Past 4,000 Years." *American Journal of Human Genetics* 107, no. 1 (July 2, 2020): 149–157.

Hammer, Michael F., Doron M. Behar, et al. "Extended Y chromosome haplotypes resolve multiple and unique lineages of the Jewish priesthood." *Human Genetics* 126, no. 5 (November 2009): 707–717.

Harney, Éadaoin, Hila May, et al. "Ancient DNA from Chalcolithic Israel reveals the role of population mixture in cultural transformation." *Nature Communications* 9 (August 20, 2018): article no. 3336. doi:10.1038/s41467-018-05649-9.

Haumann, Heiko. *A History of East European Jews*. Budapest: Central European University Press, 2002.

Hernández, Candela L., Pedro Soares, et al. "Early Holocenic and Historic mtDNA African Signatures in the Iberian Peninsula: The Andalusian Region as a Paradigm." *PLoS ONE* 10, no. 10 (October 28, 2015): e0139784. doi:10.1371/journal.pone.0139784.

Hofmanova, Zuzana, Susanne Kreutzer, et al. "Early farmers from across Europe directly descended from Neolithic Aegeans." *Proceedings of the National Academy of Sciences of the United States of America* 113, no. 25 (2016): 6886–6891.

Hundert, Gershon D. *Jews in Poland-Lithuania in the Eighteenth Century: A Genealogy of Modernity*. Berkeley, CA: University of California Press, 2004.

Jarczak, Justyna, Łukasz Grochowalski, et al. "Mitochondrial DNA variability of the Polish population." *European Journal of Human Genetics* 27 (March 21, 2019): 1304–1314.

Järve, Mari, Lehti Saag, et al. "Shifts in the Genetic Landscape of the Western Eurasian Steppe Associated with the Beginning and End of the Scythian Dominance." *Current Biology* 29, no. 14 (July 22, 2019): 2430–2441.e10. doi:10.1016/j.cub.2019.06.019.

Jasnowski, Paweł. "The Failure of the Integration of Galician Jews According to Lvov's *Ojczyzna* (1881–1892)." *Scripta Judaica Cracoviensia* 13 (2015): 55–65.

Jeong, Choongwon, Ke Wang, et al. "A Dynamic 6,000-Year Genetic History of Eurasia's Eastern Steppe." *Cell* 183, no. 4 (November 12, 2020): 890–904.e29. doi:10.1016/j.cell.2020.10.015.

Jones, Eppie R., Gunita Zarina, et al. "The Neolithic Transition in the Baltic Was Not Driven by Admixture with Early European Farmers." *Current Biology* 27, no. 4 (February 20, 2017): 576–582. doi:10.1016/j.cub.2016.12.060.

Juras, Anna, Maciej Chyleński, et al. "Investigating kinship of Neolithic post-LBK human remains from Krusza Zamkowa, Poland using ancient DNA." *Forensic Science International: Genetics* 26 (2017): 30–39. doi:10.1016/j.fsigen.2016.10.008.

Juras, Anna, Maciej Chyleński, et al. "Mitochondrial genomes reveal an east to west cline of steppe ancestry in Corded Ware populations." *Scientific Reports* 8 (August 2, 2018): article no. 11603. doi:10.1038/s41598-018-29914-5.

Juras, Anna, Maja Krzewińska, et al. "Diverse origin of mitochondrial lineages in Iron Age Black Sea Scythians." *Scientific Reports* 7 (March 7, 2017): article no. 43950. doi:10.1038/srep43950.

Juras, Anna, Miroslawa Dabert, et al. "Ancient DNA Reveals Matrilineal Continuity in Present-Day Poland over the Last Two Millennia." *PLoS ONE* 9, no. 10 (October 22, 2014): e110839. doi:10.1371/journal.pone.0110839.

Juras, Anna, Przemysław Makarowicz, et al. "Mitochondrial genomes from Bronze Age Poland reveal genetic continuity from the Late Neolithic and additional genetic affinities with the steppe populations." *American Journal of Physical Anthropology* 172, no. 2 (June 2020): 176–188.

Karpati, Mazal, Leah Peleg, et al. "A novel mutation in the HEXA gene specific to Tay-Sachs disease carriers of Jewish Iraqi origin." *Clinical Genetics* 57, no. 5 (May 2000): 398–400.

Keyser, Christine, Caroline Bouakaze, et al. "Ancient DNA provides new insights into the history of south Siberian Kurgan people." *Human Genetics* 126 (2009): 395–410.

Kiiskilä, Jukka, Jukka S. Moilanen, et al. "Analysis of functional variants in mitochondrial DNA of Finnish athletes." *BMC Genomics* 20 (October 29, 2019): article no. 784. doi:10.1186/s12864-019-6171-6.

Kizilov, Mikhail. *The Karaites of Galicia: An Ethnoreligious Minority among the Ashkenazim, the Turks, and the Slavs, 1772–1945.* Leiden: Brill, 2009.

Kizilov, Mikhail. "The Krymchaks: Current State of the Community." In *Euro-Asian Jewish Yearbook 5768 (2007/2008)*, edited by Mikhail Chlenov, et al., 63–89. Moscow: Pallada, 2009.

Kizilov, Mikhail. *The Sons of Scripture: The Karaites in Poland and Lithuania in the Twentieth Century.* Berlin: De Gruyter Open, 2015.

Klunk, Jennifer, Ana T. Duggan, et al. "Genetic resiliency and the Black Death: No apparent loss of mitogenomic diversity due to the Black Death in medieval London and Denmark." *American Journal of Physical Anthropology* 169, no. 2 (June 2019): 240–252.

Knipper, Corina, Alissa Mittnik, et al. "Female exogamy and gene pool diversification at the transition from the Final Neolithic to the Early Bronze Age in central Europe." *Proceedings of the National Academy of Sciences of the United States of America* 114, no. 38 (September 19, 2017): 10083–10088.

Knipper, Corina, Matthias Fragata, et al. "A distinct section of the Early Bronze Age society? Stable isotope investigations of burials in settlement pits and multiple inhumations of the Únětice culture in central Germany." *American Journal of Physical Anthropology* 159, no. 3 (March 2016): 496–516.

Kornienko, Igor V., Tatiana G. Faleeva, et al. "Y-Chromosome Haplogroup Diversity in Khazar Burials from Southern Russia." *Russian Journal of Genetics* 57, no. 4 (April 2021): 477–488.

Krzewińska, Maja, Anna Kjellström, et al. "Genomic and Strontium Isotope Variation Reveal Immigration Patterns in a Viking Age Town." *Current Biology* 28 (September 10, 2018): 2730–2738. doi:10.1016/j.cub.2018.06.053.

Kulik, Alexander, and Judith Kalik. "The Beginnings of Polish Jewry: Reevaluating the Evidence for the Eleventh to Fourteenth Centuries." *Zeitschrift für Ostmitteleuropa-Forschung* 70, no. 2 (2021): 139–185.

Kutanan, Wibhu, Jatupol Kampuansai, et al. "Complete mitochondrial genomes of Thai and Lao populations indicate an ancient origin of Austroasiatic groups and demic diffusion in the spread of Tai-Kadai languages." *Human Genetics* 136, no. 1 (January 2017): 85–98.

Lazaridis, Iosif, Dani Nadel, et al. "Genomic insights into the origin of farming in the ancient Near East." *Nature* 536, no. 7617 (July 25, 2016): 419–424.

Li, Shengting, Soren Besenbacher, et al. "Variation and association to diabetes in 2000 full mtDNA sequences mined from an exome study in a Danish population." *European Journal of Human Genetics* 22, no. 8 (August 2014): 1040–1045.

Linderholm, Anna, Gülşah Merve Kılınç, et al. "Corded Ware cultural complexity uncovered using genomic and isotopic analysis from south-eastern Poland." *Scientific Reports* 10 (April 14, 2020): article no. 6885. doi:10.1038/s41598-020-63138-w.

Lipson, Mark, Anna Szécsényi-Nagy, et al. "Parallel palaeogenomic transects reveal complex genetic history of early European farmers." *Nature* 551, no. 7680 (November 16, 2017): 368–372.

Loosdrecht, Mariekevan de, Abdeljalil Bouzouggar, et al. "Pleistocene North African genomes link Near Eastern and sub-Saharan African human populations." *Science* 360, no. 6388 (May 4, 2018): 548–552.

Maár, Kitti, Gergely I. B. Varga, et al. "Maternal Lineages from 10–11th Century Commoner Cemeteries of the Carpathian Basin." *Genes (Basel)* 12, no. 3 (March 23, 2021): article no. 460. doi:10.3390/genes12030460.

Maciejko, Paweł. *The Mixed Multitude: Jacob Frank and the Frankist Movement, 1755–1816.* Philadelphia: University of Pennsylvania Press, 2011.

Malmström, Helena, Anna Linderholm, et al. "Ancient mitochondrial DNA from the northern fringe of the Neolithic farming expansion in Europe sheds light on the dispersion process." *Philosophical Transactions of the Royal Society B: Biological Sciences* 370, no. 1660 (January 19, 2015): article no. 20130373. doi:10.1098/rstb.2013.0373.

Malmström, Helena, Torsten Günther, et al. "The genomic ancestry of the Scandinavian Battle Axe Culture people and their relation to the broader Corded Ware horizon." *Proceedings of the Royal Society B: Biological Sciences* 286, no. 1912 (October 9, 2019): article no. 20191528. doi:10.1098/rspb.2019.1528.

Malyarchuk, Boris A., Andrey Litvinov, et al. "Mitogenomic diversity in Russians and Poles." *Forensic Science International: Genetics* 30 (September 2017): 51–56. doi:10.1016/j.fsigen.2017.06.003.

Malyarchuk, Boris A., Miroslava Derenko, et al. "Mitogenomic diversity in Tatars from the Volga-Ural region of Russia." *Molecular Biology and Evolution* 27, no. 10 (2010): 2220–2226.

Malyarchuk, Boris A., Miroslava Derenko, et al. "The peopling of Europe from the mitochondrial haplogroup U5 perspective." *PLoS ONE* 5, no. 4 (April 21, 2010): e10285. doi:10.1371/journal.pone.0010285.

Malyarchuk, Boris A., Miroslava Derenko, et al. "Whole mitochondrial genome diversity in two Hungarian populations." *Molecular Genetics and Genomics* 293, no. 5 (October 2018): 1255–1263.

Marcus, Ivan G. "Why Did Medieval Northern French Jewry (Ṣarfat) Disappear?" In *Jews, Christians and Muslims in Medieval and Early Modern Times*, edited by Arnold E. Franklin, et al., 99–117. Leiden: Brill, 2014.

Marcus, Joseph H., Cosimo Posth, et al. "Genetic history from the Middle Neolithic to present on the Mediterranean island of Sardinia." *Nature Communications* 11 (February 24, 2020): article no. 939. doi:10.1038/s41467-020-14523-6.

Margaryan, Ashot, Miroslava Derenko, et al. "Eight Millennia of Matrilineal Genetic Continuity in the South Caucasus." *Current Biology* 27, no. 13 (July 10, 2017): 2023–2028.e7. doi:10.1016/j.cub.2017.05.087.

Margaryan, Ashot, Daniel J. Lawson, et al. "Population genomics of the Viking world." *Nature* 585, no. 7825 (September 16, 2020): 390–396.

Martiniano, Rui, Anwen Caffell, et al. "Genomic signals of migration and continuity in Britain before the Anglo-Saxons." *Nature Communications* 7 (January 19, 2016): article no. 10326. doi:10.1038/ncomms10326.

Mathieson, Iain, Iosif Lazaridis, et al. "Genome-wide patterns of selection in 230 ancient Eurasians." *Nature* 528, no. 7583 (November 23, 2015): 499–503.

Mathieson, Iain, Songül Alpaslan-Roodenberg, et al. "The genomic history of southeastern Europe." *Nature* 555, no. 7695 (March 8, 2018): 197–203.

Matisoo-Smith, Elizabeth, Anna L. Gosling, et al. "Ancient mitogenomes of Phoenicians from Sardinia and Lebanon: A story of settlement, integration, and female mobility." *PLoS ONE* 13, no. 1 (January 10, 2018): e0190169. doi:10.1371/journal.pone.0190169.

McDowell, G. A., E. H. Mules, et al. "The presence of two different infantile Tay–Sachs disease mutations in a Cajun population." *American Journal of Human Genetics* 51, no. 5 (November 1992): 1071–1077.

Mielnik-Sikorska, Marta, Patrycja Daca, et al. "The History of Slavs Inferred from Complete Mitochondrial Genome Sequences." *PLoS ONE* 8, no. 1 (January 14, 2013): e54360. doi:10.1371/journal.pone.0054360.

Mikheyev, Alexander S., Lijun Qiu, et al. "Diverse genetic origins of medieval steppe nomad conquerors." bioRxiv, December 16, 2019. doi:10.1101/2019.12.15.876912. Accessed October 3, 2021. https://www.biorxiv.org/content/10.1101/2019.12.15.876912v1

Mittnik, Alissa, Chuan-Chao Wang, et al. "The genetic prehistory of the Baltic Sea region." *Nature Communications* 9 (January 30, 2018): article no. 442. doi:10.1038/s41467-018-02825-9.

Modi, Alessandra, Desislava Nesheva, et al. "Ancient human mitochondrial genomes from Bronze Age Bulgaria: new insights into the genetic history of Thracians." *Scientific Reports* 9 (April 1, 2019): article no. 5412. doi:10.1038/s41598-019-41945-0.

Modi, Alessandra, Hovirag Lancioni, et al. "The mitogenome portrait of Umbria in Central Italy as depicted by contemporary inhabitants and pre-Roman remains." *Scientific Reports* 10 (July 1, 2020): article no. 10700. doi:10.1038/s41598-020-67445-0.

Müldner, Gundula, Carolyn Chenery, and Hella Eckardt. "The 'Headless Romans': Multi-isotope investigations of an unusual burial ground from Roman Britain." *Journal of Archaeological Science* 38, no. 2 (February 2011): 280–290.

Musilová, Eliška, Verónica Fernandes, et al. "Population History of the Red Sea—Genetic Exchanges Between the Arabian Peninsula and East Africa Signaled in the Mitochondrial DNA HV1 Haplogroup." *American Journal of Physical Anthropology* 145 (2011): 592–598.

Nagy, Péter L., Judit Olasz, et al. "Determination of the phylogenetic origins of the Árpád Dynasty based on Y chromosome sequencing of Béla the Third." *European Journal of Human Genetics* 29, no. 1 (January 2021): 164–172.

Narasimhan, Vagheesh M., Nick Patterson, et al. "The formation of human populations in South and Central Asia." *Science* 365, no. 6457 (September 6, 2019): eaat7487.

Neparáczki, Endre, Klaudia Kocsy, et al. "Revising mtDNA haplotypes of the ancient Hungarian conquerors with next generation sequencing." *PLoS ONE* 12, no. 4 (April 19, 2017): e0174886. doi:10.1371/journal.pone.0174886.

Neparáczki, Endre, Zoltán Maróti, et al. "Mitogenomic data indicate admixture components of Central-Inner Asian and Srubnaya origin in the conquering Hungarians." *PLoS ONE* 13, no. 10 (October 18, 2018): e0205920. doi:10.1371/journal.pone.0205920.

Nogueiro, Inês, João C. Teixeira, et al. "Echoes from Sepharad: signatures on the maternal gene pool of crypto-Jewish descendants." *European Journal of Human Genetics* 23, no. 5 (May 2015): 693-699.

Nogueiro, Inês, João C. Teixeira, et al. "Portuguese crypto-Jews: the genetic heritage of a complex history." *Frontiers in Genetics* 6 (February 2, 2015): article no. 12. doi:10.3389/fgene.2015.00012.

Non, Amy L., Ali Al-Meeri, et al. "Mitochondrial DNA reveals distinct evolutionary histories for Jewish populations in Yemen and Ethiopia." *American Journal of Physical Anthropology* 144, no. 1 (January 2011): 1-10.

Novak, Mario, Iñigo Olalde, et al. "Genome-wide analysis of nearly all the victims of a 6200 year old massacre." *PLoS ONE* 16, no. 3 (March 10, 2021): e0247332. doi:10.1371/journal.pone.0247332.

Olalde, Iñigo, Selina Brace, et al. "The Beaker phenomenon and the genomic transformation of northwest Europe." *Nature* 555, no. 7695 (March 8, 2018): 190-196.

Olalde, Iñigo, Swapan Mallick, et al. "The genomic history of the Iberian Peninsula over the past 8000 years." *Science* 363, no. 6432 (2019): 1230-1234.

Olivieri, Anna, Carlo Sidore, et al. "Mitogenome Diversity in Sardinians: A Genetic Window onto an Island's Past." *Molecular Biology and Evolution* 34, no. 5 (May 2017): 1230-1239.

Olivieri, Anna, Maria Pala, et al. "Mitogenomes from Two Uncommon Haplogroups Mark Late Glacial/Postglacial Expansions from the Near East and Neolithic Dispersals within Europe." *PLoS ONE* 8, no. 7 (July 31, 2013): e70492. doi:10.1371/journal.pone.0070492.

Ostrer, Harry. *Legacy: A Genetic History of the Jewish People*. New York: Oxford University Press USA, 2012.

Översti, Sanni, Kerttu Majander, et al. "Human mitochondrial DNA lineages in Iron-Age Fennoscandia suggest incipient admixture and eastern introduction of farming-related maternal ancestry." *Scientific Reports* 9 (November 15, 2019): article no. 16883. doi:10.1038/s41598-019-51045-8.

Pala, Maria, Anna Olivieri, et al. "Mitochondrial DNA Signals of Late Glacial Recolonization of Europe from Near Eastern Refugia." *American Journal of Human Genetics* 90, no. 5 (May 4, 2012): 915-924.

Patterson, Nick, Michael Isakov, et al. "Large-scale migration into Britain during the Middle to Late Bronze Age." *Nature* 601, no. 7894 (January 27, 2022): 588-594.

Peng, Min-Sheng, Weifang Xu, et al. "Mitochondrial genomes uncover the maternal history of the Pamir populations." *European Journal of Human Genetics* 26, no. 1 (January 2018): 124-136.

Pereira, Luísa, Martin Richards, et al. "High-resolution mtDNA evidence for the late-glacial resettlement of Europe from an Iberian refugium." *Genome Research* 15, no. 1 (January 2005): 19–24.

Pereira, Luísa, Nuno M. Silva, et al. "Population expansion in the North African late Pleistocene signalled by mitochondrial DNA haplogroup U6." *BMC Evolutionary Biology* 10 (December 21, 2010): article no. 390. doi:10.1186/1471-2148-10-390.

Piotrowska-Nowak, Agnieszka, Ewa Kosior-Jarecka, et al. "Investigation of whole mitochondrial genome variation in normal tension glaucoma." *Experimental Eye Research* 178 (January 2019): 186–197.

Piotrowska-Nowak, Agnieszka, Joanna L. Elson, et al. "New mtDNA Association Model, MutPred Variant Load, Suggests Individuals With Multiple Mildly Deleterious mtDNA Variants Are More Likely to Suffer From Atherosclerosis." *Frontiers in Genetics* 9 (2018): article no. 702. doi:10.3389/fgene.2018.00702.

Raule, Nicola, Federica Sevini, et al. "The co-occurrence of mtDNA mutations on different oxidative phosphorylation subunits, not detected by haplogroup analysis, affects human longevity and is population specific." *Aging Cell* 13, no. 3 (June 2014): 401–407.

Reiner, Rami. "Tough are Gerim: Conversion to Judaism in Medieval Europe." *Havruta: A Journal of Jewish Conversation* 1 (Spring 2008): 54–63.

Réthelyi, Mari. "Hungarian Jewish Stories of Origin: Samuel Kohn, the Khazar Connection and the Conquest of Hungary." *Hungarian Cultural Studies: e-Journal of the American Hungarian Educators Association* 14 (2021): 52–64. doi:10.5195/ahea.2021.427.

Rootsi, Siiri, Doron M. Behar, et al. "Phylogenetic Applications of Whole Y-Chromosome Sequences and the Near Eastern Origin of Ashkenazi Levites." *Nature Communications* 4 (December 17, 2013): article no. 2928. doi:10.1038/ncomms3928.

Rosenthal, Herman. "Lithuania." In *The Jewish Encyclopedia*, edited by Isidore Singer, vol. 8, 118–130. New York: Funk & Wagnalls, 1904.

Rosman, Moshe. "Poland before 1795." In *The YIVO Encyclopedia of Jews in Eastern Europe*, edited by Gershon D. Hundert, vol. 2, 1381–1389. New Haven, CT: Yale University Press, 2008.

Rubinstein, Arthur. *My Young Years.* New York: Alfred A. Knopf, 1973.

Rusu, Ioana, Alessandra Modi, et al. "Mitochondrial ancestry of medieval individuals carelessly interred in a multiple burial from southeastern Romania." *Scientific Reports* 9 (January 30, 2019): article no. 961. doi:10.1038/s41598-018-37760-8.

Rusu, Ioana, Alessandra Modi, et al. "Maternal DNA lineages at the gate of Europe in the 10th century AD." *PLoS ONE* 13, no. 3 (March 14, 2020): e0193578. doi:10.1371/journal.pone.0193578.

Saag, Lehti, Liivi Varul, et al. "Extensive Farming in Estonia Started through a Sex-Biased Migration from the Steppe." *Current Biology* 27, no. 14 (July 24, 2017): 2185–2193.e6. doi:10.1016/j.cub.2017.06.022.

Saag, Lehti, Margot Laneman, et al. "The Arrival of Siberian Ancestry Connecting the Eastern Baltic to Uralic Speakers further East." *Current Biology* 29, no. 10 (May 20, 2019): 1701–1711.e16. doi:10.1016/j.cub.2019.04.026.

Saag, Lehti, Sergey V. Vasilyev, et al. "Genetic ancestry changes in Stone to Bronze Age transition in the East European plain." *Science Advances* 7, no. 4 (January 20, 2021): eabd6535. doi:10.1126/sciadv.abd6535.

Sahakyan, Hovhannes, Baharak H. Kashani, et al. "Origin and spread of human mitochondrial DNA haplogroup U7." *Scientific Reports* 7 (April 7, 2017): article no. 46044. doi:10.1038/srep46044.

Sánchez-Quinto, Federico, Helena Malmström, et al. "Megalithic tombs in western and northern Neolithic Europe were linked to a kindred society." *Proceedings of the National Academy of Sciences of the United States of America* 116, no. 19 (May 7, 2019): 9469–9474.

Santoro, Aurelia, Valentina Balbi, et al. "Evidence for Sub-Haplogroup H5 of Mitochondrial DNA as a Risk Factor for Late Onset Alzheimer's Disease." *PLoS ONE* 5, no. 8 (August 6, 2010): e12037. doi:10.1371/journal.pone.0012037.

Saupe, Tina, Francesco Montinaro, et al. "Ancient genomes reveal structural shifts after the arrival of Steppe-related ancestry in the Italian peninsula." *Current Biology* 31, no. 9 (May 10, 2021): S0960–9822(21)00535–2. doi:10.1016/j.cub.2021.04.022.

Schönberg, Anna, Christoph Theunert, et al. "High-throughput sequencing of complete human mtDNA genomes from the Caucasus and West Asia: high diversity and demographic inferences." *European Journal of Human Genetics* 19 (April 13, 2011): 988–994.

Schuenemann, Verena J., Alexander Peltzer, et al. "Ancient Egyptian mummy genomes suggest an increase of Sub-Saharan African ancestry in post-Roman periods." *Nature Communications* 8 (May 30, 2017): article no. 15694. doi:10.1038/ncomms15694.

Scorrano, Gabriele, Andrea Finocchio, et al. "The Genetic Landscape of Serbian Populations through Mitochondrial DNA Sequencing and Non-Recombining Region of the Y Chromosome Microsatellites." *Collegium Antropologicum* 41, no. 3 (2017): 275–296.

Sécher, Bernard, Rosa Fregel, et al. "The history of the North African mitochondrial DNA haplogroup U6 gene flow into the African, Eurasian and American continents." *BMC Evolutionary Biology* 14 (May 19, 2014): article no. 109. doi:10.1186/1471-2148-14-109.

Semenov, Alexander S., and Vladimir V. Bulat. "Ancient Paleo-DNA of Pre-Copper Age North-Eastern Europe: Establishing the Migration Traces of R1a1 Y-DNA Haplogroup." *European Journal of Molecular Biotechnology* 11, no. 1 (2016): 40–54. doi:10.13187/ejmb.2016.11.40.

Serra-Vidal, Gerard, Marcel Lucas-Sanchez, et al. "Heterogeneity in Palaeolithic Population Continuity and Neolithic Expansion in North Africa." *Current Biology* 29, no. 22 (November 18, 2019): 3953–3959.e4. doi:10.1016/j.cub.2019.09.050.

Shamoon-Pour, Michel, Mian Li, et al. "Rare human mitochondrial HV lineages spread from the Near East and Caucasus during post-LGM and Neolithic expansions." *Scientific Reports* 9 (October 14, 2019): article no. 14751. doi:10.1038/s41598-019-48596-1.

Shapira, Dan. "Beginnings of the Karaites of the Crimea Prior to the Early Sixteenth Century." In *A Guide to Karaite Studies: An Introduction to the Literary Sources of Medieval and Modern Karaite Judaism*, edited by Meira Polliack, 709–728. Leiden: Brill, 2003.

Shen, Peidong, Tal Lavi, et al. "Reconstruction of Patrilineages and Matrilineages of Samaritans and Other Israeli Populations From Y-Chromosome and Mitochondrial DNA Sequence Variation." *Human Mutation* 24, no. 3 (September 2004): 248–260.

Shlush, Liran I., Doron M. Behar, et al. "The Druze: a population genetic refugium of the Near East." *PLoS ONE* 3, no. 5 (May 7, 2008): e2105. doi:10.1371/journal.pone.0002105.

Silva, Marina, Farida Alshamali, et al. "60,000 years of interactions between Central and Eastern Africa documented by major African mitochondrial haplogroup L2." *Scientific Reports* 5 (July 27, 2015): article no. 12526. doi:10.1038/srep12526.

Silva, Marina, Gonzalo Oteo-García, et al. "Biomolecular insights into North African-related ancestry, mobility and diet in eleventh-century Al-Andalus." *Scientific Reports* 11 (September 13, 2021): article no. 18121. doi:10.1038/s41598-021-95996-3.

Skourtanioti, Eirini, Yilmaz S. Erdal, et al. "Genomic History of Neolithic to Bronze Age Anatolia, Northern Levant, and Southern Caucasus." *Cell* 181, no. 5 (May 28, 2020): 1158–1175.e28. doi:10.1016/j.cell.2020.04.044.

Stanizai, Zaman. "Are Pashtuns the Lost Tribe of Israel?" Cambridge Open Engage, June 28, 2021. doi:10.33774/coe-2020-vntk7-v6. Accessed January 31, 2022. https://www.cambridge.org/engage/coe/article-details/60d49c27c62295e4ef1ade46

Stolarek, Ireneusz, Anna Juras, et al. "A mosaic genetic structure of the human population living in the South Baltic region during the Iron Age." *Scientific Reports* 8 (February 6, 2018): article no. 2455. doi:10.1038/s41598-018-20705-6.

Stolarek, Ireneusz, Luiza Handschuh, et al. "Goth migration induced changes in the matrilineal genetic structure of the central-east European population." *Scientific Reports* 9 (May 1, 2019): article no. 6737. doi:10.1038/s41598-019-43183-w.

Straub, David. "Jews in Azerbaijan." In *Encyclopedia of the Jewish Diaspora: Origins, Experiences, and Culture*, edited by M. Avrum Ehrlich, vol. 1, 1114–1119. Santa Barbara, CA: ABC-CLIO, 2008.

Sukernik, Rem I., Natalia V. Volodko, et al. "Mitochondrial genome diversity in the Tubalar, Even, and Ulchi: contribution to prehistory of native Siberians and their affinities to Native Americans." *American Journal of Physical Anthropology* 148, no. 1 (May 2012): 123–138.

Tambets, Kristiina, Bayazit Yunusbayev, et al. "Genes reveal traces of common recent demographic history for most of the Uralic-speaking populations." *Genome Biology* 19 (September 21, 2018): article no. 139. doi:10.1186/s13059-018-1522-1.

Teter, Magda. "Jewish Conversions to Catholicism in the Polish-Lithuanian Commonwealth of the Seventeenth and Eighteenth Centuries." *Jewish History* 17, no. 3 (2003): 257–283.

Teter, Magda. *Jews and Heretics in Catholic Poland: A Beleaguered Church in the Post-Reformation Era*. Cambridge: Cambridge University Press, 2006.

Teter, Magda. "Conversion." In *The YIVO Encyclopedia of Jews in Eastern Europe*, edited by Gershon D. Hundert, vol. 1, 348–351. New Haven, CT: Yale University Press, 2008.

Tian, Jiao-Yang, Hua-Wei Wang, et al. "A Genetic Contribution from the Far East into Ashkenazi Jews via the Ancient Silk Road." *Scientific Reports* 5 (February 11, 2015): article no. 8377. doi:10.1038/srep08377.

Toaff, Ariel, and Nadia Zeldes. "Taranto." In *Encyclopaedia Judaica*, 2nd ed., edited by Fred Skolnik, vol. 19. Detroit: Macmillan Reference USA, 2008.

Unkefer, Rachel. "Disproving a Cossack Paternal Ancestry for an Ashkenazic Lineage." *Avotaynu* 36, no. 3 (Fall 2020): 46–51.

Unterländer, Martina, Friso Palstra, et al. "Ancestry and demography and descendants of Iron Age nomads of the Eurasian Steppe." *Nature Communications* 8 (March 3, 2017): article no. 14615. doi:10.1038/ncomms14615.

Vai, Stefania, Andrea Brunelli, et al. "A genetic perspective on Longobard-Era migrations." *European Journal of Human Genetics* 27, no. 4 (April 2019): 647–656.

van Straten, Jits. *Ashkenazic Jews and the Biblical Israelites: The Early Demographic Development of East European Ashkenazis*. Berlin: De Gruyter Oldenbourg, 2021.

Veeramah, Krishna R., Andreas Rott, et al. "Population genomic analysis of elongated skulls reveals extensive female-biased immigration in Early Medieval Bavaria." *Proceedings of the National Academy of Sciences of the United States of America* 115, no. 13 (March 27, 2018): 3494–3499.

Vyas, Deven N., Andrew Kitchen, et al. "Bayesian analyses of Yemeni mitochondrial genomes suggest multiple migration events with Africa and Western Eurasia." *American Journal of Physical Anthropology* 159, no. 3 (March 2016): 382–393.

Weinryb, Bernard D. *The Jews of Poland: A Social and Economic History of the Jewish Community in Poland from 1100 to 1800*. Philadelphia: Jewish Publication Society of America, 1973.

Wierzbieniec, Wacław. "Zamość." In *The YIVO Encyclopedia of Jews in Eastern Europe*, edited by Gershon D. Hundert, vol. 2, 2111–2112. New Haven, CT: Yale University Press, 2008.

Wilde, Sandra, Adrian Timpson, et al. "Direct evidence for positive selection of skin, hair, and eye pigmentation in Europeans during the last 5,000 y." *Proceedings of the National Academy of Sciences of the United States of America* 111, no. 13 (April 1, 2014): 4832–4837.

Xue, James, Todd Lencz, et al. "The Time and Place of European Admixture in Ashkenazic Jewish History." *PLoS Genetics* 13, no. 4 (April 4, 2017): e1006644. doi:10.1371/journal.pgen.1006644.

Yacobi, Doron, and Felice L. Bedford. "Evidence of Early Gene Flow between Ashkenazi Jews and Non-Jewish Europeans in Mitochondrial DNA Haplogroup H7." *Journal of Genetic Genealogy* 8, no. 1 (Fall 2016): 21–34. Accessed October 3, 2021. https://jogg. info/pages/vol8/yacobi2016/Evidence-of-early-gene-flow-between-Ashkenazi-Jews-and-non-Jewish-Europeans-in-mitochondrial-DNA-haplogroup-H7.pdf

Yaka, Reyhan, Ayşegül Birand, et al. "Archaeogenetics of Late Iron Age Çemialo Sırtı, Batman: Investigating maternal genetic continuity in north Mesopotamia since the Neolithic." *American Journal of Physical Anthropology* 166, no. 1 (May 2018): 196–207.

Zalloua, Pierre, Catherine J. Collins, et al. "Ancient DNA of Phoenician remains indicates discontinuity in the settlement history of Ibiza." *Scientific Reports* 8 (December 4, 2018): article no. 17567. doi:10.1038/s41598-018-35667-y.

Zhang, Xiaoming, Chunmei Li, et al. "A Matrilineal Genetic Perspective of Hanging Coffin Custom in Southern China and Northern Thailand." *iScience* 23 (April 24, 2020): 101032. doi:10.1016/j.isci.2020.101032.

Zhao, Dan, Yingying Ding, et al. "Mitochondrial Haplogroups N9 and G Are Associated with Metabolic Syndrome Among Human Immunodeficiency Virus-Infected Patients in China." *AIDS Research and Human Retroviruses* 35, no. 6 (June 2019): 536–543.

Zhao, Jing, Wurigemule Wurigemule, et al. "Genetic substructure and admixture of Mongolians and Kazakhs inferred from genome-wide array genotyping." *Annals of Human Biology* 47, no. 7–8 (December 2020): 620–628.

Index

About the Author

Kevin Alan Brook authored the peer-reviewed study "The Genetics of Crimean Karaites" in the Summer 2014 issue of *Karadeniz Araştırmaları*, based on original research, and has worked as a genetic genealogy researcher specializing in Sephardic Jewish DNA. He is also the author of the book *The Jews of Khazaria, Third Edition* (Rowman & Littlefield, 2018), the peer-reviewed article "The Origins of East European Jews" in the Spring-Summer 2003 issue of *Russian History/Histoire Russe,* and the article "Jews in Medieval Armenia" in *Encyclopedia of the Jewish Diaspora* (ABC-CLIO, 2008).

9 781644 699843